HOLT

Physics

Section Quizzes

HOLT, RINEHART AND WINSTON

A Harcourt Education Company

Orlando • **Austin** • New York • San Diego • Toronto • London

ISBN 0-03-036837-5

15 16 17 1410 15 14 13 12 11
4500316382

Contents

Assessment

Chapter Test A

The Science of Physics
MULTIPLE CHOICE

In the space provided, write the letter of the term or phrase that best completes each statement or best answers each question.

_____ **1.** Which of the following is an area of physics that studies motion and its causes?

    a. thermodynamics   **c.** quantum mechanics

    b. mechanics   **d.** optics

_____ **2.** Listening to your favorite radio station involves which area of physics?

    a. optics

    b. thermodynamics

    c. vibrations and wave phenomena

    d. relativity

_____ **3.** A baker makes a loaf of bread. Identify the area of physics that this involves.

    a. optics   **c.** mechanics

    b. thermodynamics   **d.** relativity

_____ **4.** According to the scientific method, why does a physicist make observations and collect data?

    a. to decide which parts of a problem are important

    b. to ask a question

    c. to make an interpretation

    d. to solve all problems

_____ **5.** In the steps of the scientific method, what is the next step after formulating and objectively testing hypotheses?

    a. interpreting results

    b. stating conclusions

    c. conducting experiments

    d. making observations and collecting data

_____ **6.** Why do physicists use models?

    a. to explain the complex features of simple phenomena

    b. to describe all aspects of a phenomenon

    c. to explain the basic features of complex phenomena

    d. to describe all of reality

_____ **7.** Which statement about models is *not* correct?
  a. Models describe only part of reality.
  b. Models help build hypotheses.
  c. Models help guide experimental design.
  d. Models manipulate a single variable or factor in an experiment.

_____ **8.** What two dimensions, in addition to mass, are commonly used by physicists to derive additional measurements?
  a. length and width   **c.** length and time
  b. area and mass   **d.** velocity and time

_____ **9.** The symbol mm represents a
  a. micrometer.   **c.** megameter.
  b. millimeter.   **d.** manometer.

_____**10.** The SI base unit used to measure mass is the
  a. meter.   **c.** kilogram.
  b. second.   **d.** liter.

_____**11.** The SI base unit for time is
  a. 1 day.   **c.** 1 minute.
  b. 1 hour.   **d.** 1 second.

_____**12.** A lack of precision in scientific measurements typically arises from
  a. limitations of the measuring instrument.
  b. human error.
  c. lack of calibration.
  d. too many significant figures.

_____**13.** How does a scientist reduce the frequency of human error and minimize a lack of accuracy?
  a. Take repeated measurements.
  b. Use the same method of measurement.
  c. Maintain instruments in good working order.
  d. all of the above

_____**14.** Five darts strike near the center of a target. The dart thrower is
  a. accurate.   **c.** both accurate and precise.
  b. precise.   **d.** neither accurate nor precise.

_____**15.** Calculate the following, and express the answer in scientific notation with the correct number of significant figures: $21.4 + 15 + 17.17 + 4.003$
  a. 5.7573×10^1 **c.** 5.75×10^1
  b. 5.757×10^1 **d.** 5.8×10^1

Hour	Temperature (°C)
1:00	30.0
2:00	29.0
3:00	28.0
4:00	27.5
5:00	27.0
6:00	25.0

_____**16.** A weather balloon records the temperature every hour. From the table above, the temperature

    a. increases.   **c.** remains constant.

    b. decreases.   **d.** decreases and then increases.

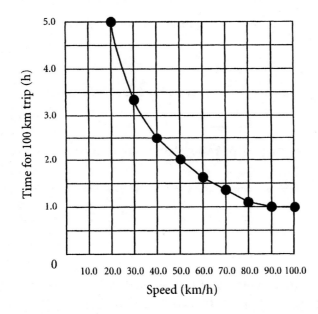

_____**17.** The time required to make a trip of 100.0 km is measured at various speeds. From the graph above, what speed will allow the trip to be made in 2 hours?

    a. 20.0 km/h   **c.** 50.0 km/h

    b. 40.0 km/h   **d.** 90.0 km/h

_____**18.** The Greek letter Δ indicates a(n)

    a. difference or change.   **c.** direct proportion.

    b. sum or total.   **d.** inverse proportion.

_____**19.** The most appropriate SI unit for measuring the length of an automobile is the

    a. micron.   **c.** meter.

    b. kilometer.   **d.** nanometer.

_____**20.** Estimate the order of magnitude of the length of a football field.

    a. 10^{-1} m   **c.** 10^4 m

    b. 10^2 m   **d.** 10^6 m

SHORT ANSWER

21. Two areas within physics are mechanics and quantum mechanics. Distinguish between these two areas.

_____

_____

22. What are the SI base units for length, mass, and time?

_____

23. Convert 92×10^3 km to decimeters using scientific notation.

_____

_____

_____

24. What must quantities have before they can be added or subtracted?

_____

_____

PROBLEM

25. Calculate the following, expressing the answer in scientific notation with the correct number of significant figures: $(8.86 + 1.0 \times 10^{-3}) \div 3.610 \times 10^{-3}$

Chapter Test B

The Science of Physics
MULTIPLE CHOICE

In the space provided, write the letter of the term or phrase that best completes each statement or best answers each question.

_____ 1. A hiker uses a compass to navigate through the woods. Identify the area of physics that this involves.
    a. thermodynamics   **c.** electromagnetism
    b. relativity   **d.** quantum mechanics

_____ 2. According to the scientific method, how does a physicist formulate and objectively test hypotheses?
    a. by defending an opinion   **c.** by experiments
    b. by interpreting graphs   **d.** by stating conclusions

_____ 3. Diagrams are *not* designed to
    a. show relationships between concepts.
    b. show setups of experiments.
    c. measure an event or a situation.
    d. label parts of a model.

_____ 4. The most appropriate SI unit for measuring the length of an automobile is the
    a. micron.   **c.** meter.
    b. kilometer.   **d.** nanometer.

_____ 5. The radius of Earth is 6 370 000 m. Express this measurement in km in scientific notation with the correct number of significant digits.
    a. 6.37×10^6 km   **c.** 637×10^3 km
    b. 6.37×10^3 km   **d.** 63.7×10^4 km

_____ 6. Three values were obtained for the mass of a metal bar: 8.83 g; 8.84 g; 8.82 g. The known mass is 10.68 g. The values are
    a. accurate.   **c.** both accurate and precise.
    b. precise.   **d.** neither accurate nor precise.

_____ 7. Calculate the following, and express the answer in scientific notation with the correct number of significant figures: $10.5 \times 8.8 \times 3.14$
    a. 2.9×10^2 **c.** 2.90×10^2
    b. 290.136   **d.** 290

_____ **8.** Calculate the following, and express the answer in scientific notation with the correct number of significant figures:

$(0.82 + 0.042)(4.4 \times 10^3)$

a. 3.8×10^3 **c.** 3.784×10^3

b. 3.78×10^3 **d.** 3784

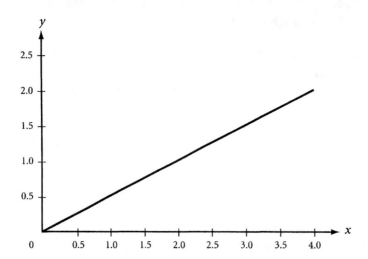

_____ **9.** Which of the following equations best describes the graph above?

a. $y = 2x$ **c.** $y = x^2$

b. $y = x$ **d.** $y = \dfrac{1}{2}x$

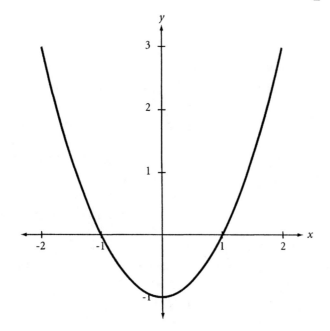

_____ **10.** Which of the following equations best describes the graph above?

a. $y = x^2 + 1$ **c.** $y = -x^2 + 1$

b. $y = x^2 - 1$ **d.** $y = -x^2 - 1$

_____**11.** Which expression has the same dimensions as an expression yielding a
value for acceleration (m/s^2)? (Δv has units of m/s.)
    a. $\Delta v/(\Delta t)^2$ **c.** $(\Delta v)^2/\Delta t$
    b. $\Delta v/(\Delta x)^2$ **d.** $(\Delta v)^2/\Delta x$

_____**12.** If the change in position Δx is related to velocity v (with units of m/s)
in the equation $\Delta x = Av$, the constant A has which dimension?
    a. m/s^2 **c.** s
    b. m   **d.** m^2

_____**13.** The sun is composed mostly of hydrogen. The mass of the sun is
2.0×10^{30} kg, and the mass of a hydrogen atom is 1.67×10^{-27} kg.
Estimate the number of atoms in the sun.
    a. 10^3 **c.** 10^{30}
    b. 10^{57} **d.** 10^{75}

SHORT ANSWER

14. If unexpected results are obtained and confirmed through repeated
experiments, why must a model or hypothesis be abandoned or revised?

_____

_____

15. How can only seven basic units serve to express almost any measured quantity?

_____

_____

16. Convert 1 μm to meters using scientific notation.

_____

_____

17. Why do calculators often exaggerate the precision of a final result?

_____

_____

18. How many significant figures does 0.050 200 mg have?

_____

_____

PROBLEM

19. The radius of Earth is 6.37×10^6 m. The average Earth-sun distance is 1.496×10^{11} m. How many Earths would fit between Earth and the sun if they are separated by their average distance? Use an order-of-magnitude calculation to estimate this number. Then, determine an exact answer and express it in scientific notation with the correct number of significant digits.

	Trial 1	Trial 2	Trial 3	Trial 4
0.0 s	20.5° C	21.3° C	20.8° C	21.0° C
5.0 s	21.0° C	22.9° C	21.4° C	21.7° C
10.0 s	21.6° C	24.1° C	22.0° C	22.3° C
15.0 s	22.2° C	26.8° C	22.7° C	22.8° C
20.0 s	23.0° C	28.2° C	23.2° C	23.3° C

20. Four trials of a chemical reaction were completed, and the change in temperature ΔT was measured every five seconds. Based on the data in the table above, answer the following questions. Are there any unexpected or unusual results? Explain your answer. What is the general relationship between temperature and time? Disregarding any trial(s) with unexpected results, express this relationship in the form of a general equation.

Name _____ Class _____ Date _____

Chapter Test A

Motion in One Dimension
MULTIPLE CHOICE

In the space provided, write the letter of the term or phrase that best completes each statement or best answers each question.

_____ **1.** What is the speed of an object at rest?

 a. 0.0 m/s   **c.** 9.8 m/s

 b. 1.0 m/s   **d.** 9.81 m/s

_____ **2.** Which of the following situations represents a negative displacement? (Assume positive position is measured vertically upward along a y-axis.)

 a. A cat stands on a tree limb.

 b. A cat jumps from the ground onto a tree limb.

 c. A cat jumps from a lower tree limb to a higher one.

 d. A cat jumps from a tree limb to the ground.

_____ **3.** Which of the following units is the SI unit of velocity?

 a. meter   **c.** meter per second

 b. meter•second   **d.** second per meter

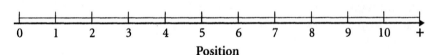

Position

_____ **4.** In the graph above, a toy car rolls from +3 m to +5 m. Which of the following statements is true?

 a. $x_f = +3$ m   **c.** $\Delta x = +3$ m

 b. $x_i = +3$ m   **d.** $v_{avg} = 3$ m/s

_____ **5.** The slope of a line drawn tangent to a point on the curve of a position versus time graph describes what concept?

 a. acceleration   **c.** instantaneous velocity

 b. displacement   **d.** position

_____ **6.** Acceleration is defined as

 a. a rate of displacement.

 b. the rate of change of displacement.

 c. the change in velocity.

 d. the rate of change of velocity.

_____ **7.** What is the SI unit of acceleration?

 a. m/s   **c.** m/s^2

 b. m^2/s **d.** $m•s^2$

| **Chapter Test A** *continued*

_____ **8.** If you know a car's acceleration, the information you must have to determine if the car's velocity is increasing is the
   a. direction of the car's initial velocity.
   b. direction of the car's acceleration.
   c. initial speed of the car.
   d. final velocity of the car.

_____ **9.** If you know the acceleration of a car and its initial velocity, you can predict which of the following?
   a. the direction of the car's final velocity
   b. the magnitude of the car's final velocity
   c. the displacement of the car
   d. all of the above

_____ **10.** When a car's velocity is positive and its acceleration is negative, what is happening to the car's motion?
   a. The car slows down.
   b. The car speeds up.
   c. The car travels at constant speed.
   d. The car remains at rest.

_____ **11.** The graph at right describes the motion of a ball. At what point does the ball have an instantaneous velocity of zero?
   a. A
   b. B
   c. C
   d. D

_____ **12.** The motion of a ball on an inclined plane is described by the equation $\Delta x = 1/2a(\Delta t)^2$. This statement implies which of the following quantities has a value of zero?
   a. x_i **c.** v_i
   b. x_f **d.** t_f

_____ **13.** Acceleration due to gravity is also called
   a. negative velocity.   **c.** free-fall acceleration.
   b. displacement.   **d.** instantaneous velocity.

_____ **14.** When there is no air resistance, objects of different masses dropped from rest
   a. fall with equal accelerations and with equal displacements.
   b. fall with different accelerations and with different displacements.
   c. fall with equal accelerations and with different displacements.
   d. fall with different accelerations and with equal displacements.

_____**15.** Objects that are falling toward Earth in free fall move
      a. faster and faster.   **c.** at a constant velocity.
      b. slower and slower.   **d.** slower then faster.

_____**16.** Which would hit the ground first if dropped from the same height in a
      vacuum—a feather or a metal bolt?
      a. the feather
      b. the metal bolt
      c. They would hit the ground at the same time.
      d. They would be suspended in a vacuum.

SHORT ANSWER

17. What is the name of the length of the straight line drawn from an object's
initial position to the object's final position?

_____

_____

18. Construct a graph of position versus
time for the motion of a dog, using
the data in the table at right. Explain
how the graph indicates that the dog
is moving at a constant speed.

Time (s)	Displacement (m)
0.0	0.0
2.0	1.0
4.0	2.0
6.0	3.0
8.0	4.0
10.0	5.0

_____

_____

PROBLEM

19. A horse trots past a fencepost located 12 m to the left of a gatepost. It then passes another fencepost located 24 m to the right of the gatepost 11 s later. What is the average velocity of the horse?

20. A rock is thrown downward from the top of a cliff with an initial speed of 12 m/s. If the rock hits the ground after 2.0 s, what is the height of the cliff? (Disregard air resistance. $a = -g = -9.81 \text{ m/s}^2$.)

Name _____ Class _____ Date _____

Chapter Test B

Motion in One Dimension
MULTIPLE CHOICE

In the space provided, write the letter of the term or phrase that best completes each statement or best answers each question.

_____ **1.** Which of the following is the equation for average velocity?

    a. $v_{avg} = \dfrac{\Delta x}{\Delta t}$

    b. $v_{avg} = \dfrac{\Delta t}{\Delta x}$

    c. $v_{avg} = \Delta x \Delta t$

    d. $v_{avg} = \dfrac{v_i - v_f}{2}$

Use the graph below to answer questions 2–4.

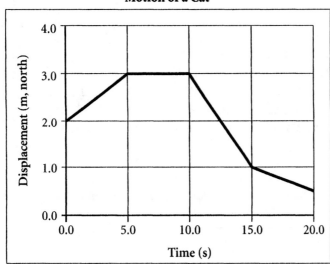

Motion of a Cat

_____ **2.** During which interval is the cat at rest?

    a. 0.0–5.0 s

    b. 5.0–10.0 s

    c. 10.0–15.0 s

    d. 15.0–20.0 s

_____ **3.** The cat has the fastest speed during which interval?

    a. 0.0–5.0 s

    b. 5.0–10.0 s

    c. 10.0–15.0 s

    d. 15.0–20.0 s

_____ **4.** During which interval does the cat have the greatest positive velocity?

    a. 0.0–5.0 s

    b. 5.0–10.0 s

    c. 10.0–15.0 s

    d. 15.0–20.0 s

_____ **5.** Which of the following is the equation for acceleration?

    a. $a = \dfrac{\Delta t}{\Delta v}$

    b. $a = \dfrac{\Delta v}{\Delta t}$

    c. $a = \Delta v \Delta t$

    d. $a = \dfrac{v_i - v_f}{t_i - t_f}$

_____ **6.** When a car's velocity is negative and its acceleration is negative, what is happening to the car's motion?
   a. The car slows down.
   b. The car speeds up.
   c. The car travels at constant speed.
   d. The car remains at rest.

_____ **7.** What does the graph on the right illustrate about acceleration?
   a. The acceleration varies.
   b. The acceleration is zero.
   c. The acceleration is constant.
   d. The acceleration increases then becomes constant.

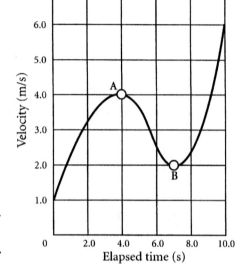

_____ **8.** In the graph on the right, how does the acceleration at A compare with the acceleration at B?
   a. The acceleration at A is positive and less than the acceleration at B.
   b. The acceleration at B is positive and less than the acceleration at A.
   c. The accelerations at A and B are each zero.
   d. The accelerations at A and B cannot be determined.

_____ **9.** Which of the following line segments on a velocity versus time graph is physically impossible?
   a. horizontal line
   b. straight line with positive slope
   c. straight line with negative slope
   d. vertical line

_____ **10.** In the graph at the right, at what point is the speed of the ball equal to its speed at B?
   a. A
   b. C
   c. D
   d. none of the above

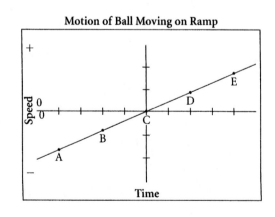

Motion of Ball Moving on Ramp

_____11. A baseball catcher throws a ball vertically upward and catches it in the same spot as it returns to the mitt. At what point in the ball's path does it experience zero velocity and nonzero acceleration at the same time?
   a. midway on the way up
   b. at the top of its path
   c. the instant it leaves the catcher's hand
   d. the instant before it arrives in the catcher's mitt

SHORT ANSWER

12. Distinguish between the displacement of a traveler who takes a train from New York to Boston and the displacement of a traveler who flies from Boston to New York. Be sure to compare the magnitudes of the displacements.

_____

_____

_____

13. If a runner moves from a positive position to a negative position, explain if the runner's displacement is negative or positive.

_____

_____

_____

14. Explain how a dog that has moved can have a displacement of zero.

_____

_____

15. Why is the direction of free-fall acceleration usually negative?

_____

_____

PROBLEM

16. A biker travels at an average speed of 18 km/h along a 0.30-km straight segment of a bike path. How much time does the biker take to travel this segment?

17. A hiker travels south along a straight path for 1.5 h with an average speed of 0.75 km/h and then travels north for 2.5 h with an average speed of 0.90 km/h. What is the hiker's displacement for the total trip?

18. A skater glides off a frozen pond onto a patch of ground at a speed of 1.8 m/s. Here she is slowed at a constant rate of 3.00 m/s^2. How fast is the skater moving when she has slid 0.37 m across the ground?

19. Human reaction time is usually about 0.20 s. If your lab partner holds a ruler between your finger and thumb and releases it without warning, how far can you expect the ruler to fall before you catch it? (Disregard air resistance. $a = -g = -9.81$ m/s^2.)

20. A pair of glasses are dropped from the top of a 32.0 m high stadium. A pen is dropped 2.00 s later. How high above the ground is the pen when the glasses hit the ground? (Disregard air resistance. $a = -g = -9.81$ m/s^2.)

Chapter Test A

Two-Dimensional Motion and Vectors
MULTIPLE CHOICE

In the space provided, write the letter of the term or phrase that best completes each statement or best answers each question.

_____ **1.** Which of the following is a physical quantity that has a magnitude but no direction?
    a. vector   **c.** resultant
    b. scalar   **d.** frame of reference

_____ **2.** Which of the following is an example of a vector quantity?
    a. velocity   **c.** volume
    b. temperature   **d.** mass

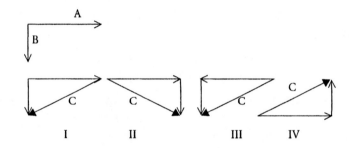

_____ **3.** In the figure above, which diagram represents the vector addition, $C = A + B$?
    a. I   **c.** III
    b. II   **d.** IV

_____ **4.** In the figure above, which diagram represents vector subtraction, $C = A - B$?
    a. I   **c.** III
    b. II   **d.** IV

_____ **5.** Multiplying or dividing vectors by scalars results in
    a. vectors.
    b. scalars.
    c. vectors if multiplied or scalars if divided.
    d. scalars if multiplied or vectors if divided.

_____ **6.** In a coordinate system, a vector is oriented at angle θ with respect to the x-axis. The x component of the vector equals the vector's magnitude multiplied by which trigonometric function?
    a. $\cos \theta$ **c.** $\sin \theta$
    b. $\cot \theta$ **d.** $\tan \theta$

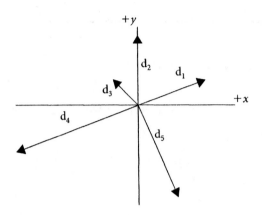

_____ **7.** How many displacement vectors shown in the figure above have horizontal components?

    a. 2   **c.** 4

    b. 3   **d.** 5

_____ **8.** Which displacement vectors shown in the figure above have vertical components that are equal?

    a. d_1 and d_2 **c.** d_2 and d_5

    b. d_1 and d_3 **d.** d_4 and d_5

_____ **9.** A hiker undergoes a displacement of d_5 as shown in the figure above. A single displacement that would return the hiker to his starting point would have which of the following sets of components?

    a. $+d_{5,\,x}$; $+d_{5,\,y}$ **c.** $-d_{5,\,x}$; $+d_{5,\,y}$

    b. $+d_{5,\,x}$; $-d_{5,\,y}$ **d.** $-d_{5,\,x}$; $-d_{5,\,y}$

_____ **10.** Which of the following is an example of projectile motion?

    a. a jet lifting off a runway

    b. a baseball being thrown

    c. dropping an aluminum can into the recycling bin

    d. a space shuttle orbiting Earth

_____ **11.** What is the path of a projectile?

    a. a wavy line

    b. a parabola

    c. a hyperbola

    d. Projectiles do not follow a predictable path.

_____ **12.** Which of the following exhibits parabolic motion?

    a. a stone thrown into a lake

    b. a space shuttle orbiting Earth

    c. a leaf falling from a tree

    d. a train moving along a flat track

_____13. Which of the following does *not* exhibit parabolic motion?
   a. a frog jumping from land into water
   b. a basketball thrown to a hoop
   c. a flat piece of paper released from a window
   d. a baseball thrown to home plate

_____14. At what point of the ball's path shown in the figure above is the
   vertical component of the ball's velocity zero?
   a. A   **c.** C
   b. B   **d.** D

_____15. A passenger on a bus moving east sees a man standing on a curb.
   From the passenger's perspective, the man appears to
   a. stand still.
   b. move west at a speed that is less than the bus's speed.
   c. move west at a speed that is equal to the bus's speed.
   d. move east at a speed that is equal to the bus's speed.

_____16. piece of chalk is dropped by a teacher walking at a speed of 1.5 m/s.
   From the teacher's perspective, the chalk appears to fall
   a. straight down.
   b. straight down and backward.
   c. straight down and forward.
   d. straight backward.

SHORT ANSWER

17. Is distance or displacement a vector quantity?

_____

_____

| Chapter Test A *continued*

18. The equation $D = \sqrt{\Delta x^2 + \Delta y^2}$ is valid only if Δx and Δy are magnitudes of vectors that have what orientation with respect to each other?

_____

_____

PROBLEM

19. A stone is thrown at an angle of 30.0° above the horizontal from the top edge of a cliff with an initial speed of 12 m/s. A stopwatch measures the stone's trajectory time from the top of the cliff to the bottom at 5.6 s. What is the height of the cliff? (Assume no air resistance and that $a_y = -g = -9.81$ m/s^2.)

20. A small airplane flies at a velocity of 145 km/h toward the south as observed by a person on the ground. The airplane pilot measures an air velocity of 172 km/h south. What is the velocity of the wind that affects the plane?

Chapter Test B

Two-Dimensional Motion and Vectors
MULTIPLE CHOICE

In the space provided, write the letter of the term or phrase that best completes each statement or best answers each question.

_____ **1.** Identify the following quantities as scalar or vector: the mass of an object, the number of leaves on a tree, wind velocity.
   a. vector, scalar, scalar
   b. scalar, scalar, vector
   c. scalar, vector, scalar
   d. vector, scalar, vector

_____ **2.** A student walks from the door of the house to the end of the driveway and realizes that he missed the bus. The student runs back to the house, traveling three times as fast. Which of the following is the correct expression for the return velocity if the initial velocity is $v_{student}$?
   a. $3v_{student}$
   b. $\frac{1}{3}v_{student}$
   c. $\frac{1}{3}v_{student}$
   d. $-3v_{student}$

_____ **3.** An ant on a picnic table travels 3.0×10^1 cm eastward, then 25 cm northward, and finally 15 cm westward. What is the magnitude of the ant's displacement relative to its original position?
   a. 70 cm
   b. 57 cm
   c. 52 cm
   d. 29 cm

_____ **4.** In a coordinate system, the magnitude of the x component of a vector and θ, the angle between the vector and x-axis, are known. The magnitude of the vector equals the x component
   a. divided by the cosine of θ.
   b. divided by the sine of θ.
   c. multiplied by the cosine of θ.
   d. multiplied by the sine of θ.

_____ **5.** Find the resultant of these two vectors: 2.00×10^2 units due east and 4.00×10^2 units 30.0° north of west.
   a. 300 units 29.8° north of west
   b. 581 units 20.1° north of east
   c. 546 units 59.3° north of west
   d. 248 units 53.9° north of west

| **Chapter Test B** *continued*

_____ **6.** In the figure at right, the magnitude of the ball's velocity is least
at location

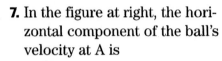

a. A.
b. B.
c. C.
d. D.

_____ **7.** In the figure at right, the horizontal component of the ball's
velocity at A is

a. zero.
b. equal to the vertical component of the ball's velocity at C.
c. equal in magnitude but opposite in direction to the horizontal
component of the ball's velocity at D.
d. equal to the horizontal component of its initial velocity.

_____ **8.** A track star in the long jump goes into the jump at 12 m/s and
launches herself at 20.0° above the horizontal. What is the magnitude
of her horizontal displacement? (Assume no air resistance and that
$a_y = -g = -9.81 \text{ m/s}^2$.)

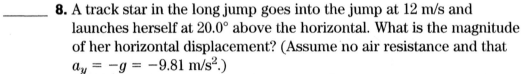

a. 4.6 m   **c.** 13 m
b. 9.2 m   **d.** 15 m

_____ **9.** A boat travels directly across a river that has a downstream current, **v.**
What is true about the perpendicular components of the boat's velocity?

a. One component equals **v**; the other component equals zero.
b. One component is perpendicular to **v**; the other component
equals **v.**
c. One component is perpendicular to **v**; the other component
equals −**v.**
d. One component is perpendicular to **v**; the other component
equals zero.

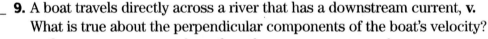

_____ **10.** A jet moving at 500.0 km/h due east is in a region where the wind is
moving at 120.0 km/h in a direction 30.00° north of east. What is the
speed of the aircraft relative to the ground?

a. 620.2 km/h   **c.** 588.7 km/h
b. 606.9 km/h   **d.** 511.3 km/h

SHORT ANSWER

11. Briefly explain the triangle (or polygon) method of addition.

_____

_____

_____

12. If the magnitude of a component vector equals the magnitude of the vector, then what is the magnitude of the other component vector?

_____

_____

13. How can you use the Pythagorean theorem to add two vectors that are not perpendicular?

_____

_____

_____

_____

14. Briefly explain why a basketball being thrown toward a hoop is considered projectile motion.

_____

_____

PROBLEM

15. A cave explorer travels 3.0 m eastward, then 2.5 m northward, and finally 15.0 m westward. Use the graphical method to find the magnitude of the net displacement.

▌Chapter Test B *continued*

16. A dog walks 28 steps north and then walks 55 steps west to bury a bone. If the dog walks back to the starting point in a straight line, how many steps will the dog take? Use the graphical method to find the magnitude of the net displacement.

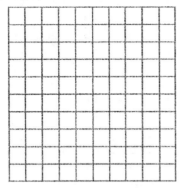

17. A quarterback takes the ball from the line of scrimmage and runs backward for 1.0×10^1 m then sideways parallel to the line of scrimmage for 15 m. The ball is thrown forward 5.0×10^1 m perpendicular to the line of scrimmage. The receiver is tackled immediately. How far is the football displaced from its original position?

18. Vector **A** is 3.2 m in length and points along the positive y-axis. Vector **B** is 4.6 m in length and points along a direction 195° counterclockwise from the positive x-axis. What is the magnitude of the resultant when vectors **A** and **B** are added?

19. A model rocket flies horizontally off the edge of a cliff at a velocity of 50.0 m/s. If the canyon below is 100.0 m deep, how far from the edge of the cliff does the model rocket land? ($a_y = -g = -9.81$ m/s^2)

20. A boat moves at 10.0 m/s relative to the water. If the boat is in a river where the current is 2.00 m/s, how long does it take the boat to make a complete round trip of 1000.0 m upstream followed by 1000.0 m downstream?

Assessment

Chapter Test A

Forces and the Laws of Motion
MULTIPLE CHOICE

In the space provided, write the letter of the term or phrase that best completes each statement or best answers each question.

_____ **1.** Which of the following is the cause of an acceleration?
    a. speed   **c.** force
    b. inertia   **d.** velocity

_____ **2.** What causes a moving object to change direction?
    a. acceleration   **c.** inertia
    b. velocity   **d.** force

_____ **3.** Which of the following forces exists between objects even in the absence of direct physical contact?
    a. frictional force   **c.** contact force
    b. fundamental force   **d.** field force

_____ **4.** A newton is equivalent to which of the following quantities?
    a. kg   **c.** $kg \bullet m/s^2$
    b. $kg \bullet m/s$ **d.** $kg \bullet (m/s)^2$

_____ **5.** The length of a force vector represents the
    a. cause of the force.
    b. direction of the force.
    c. magnitude of the force.
    d. type of force.

_____ **6.** A free-body diagram represents all of the following *except*
    a. the object.
    b. forces as vectors.
    c. forces exerted by the object.
    d. forces exerted on the object.

_____ **7.** In the free-body diagram shown to the right, which of the following is the gravitational force acting on the car?
    a. 5800 N   **c.** 14 700 N
    b. 775 N   **d.** 13 690 N

_____ **8.** Which of the following is the tendency of an object to maintain its
state of motion?
  a. acceleration   **c.** force
  b. inertia   **d.** velocity

_____ **9.** A crate is released on a frictionless plank inclined at angle θ with
respect to the horizontal. Which of the following relationships is true?
(Assume that the x-axis is parallel to the surface of the incline.)
  a. $F_y = F_g$ **c.** $F_y = F_x$
  b. $F_x = 0$ **d.** none of the above

_____ **10.** A car goes forward along a level road at constant velocity. The additional force needed to bring the car into equilibrium is
  a. greater than the normal force times the coefficient of static friction.
  b. equal to the normal force times the coefficient of static friction.
  c. the normal force times the coefficient of kinetic friction.
  d. zero.

_____ **11.** If a nonzero net force is acting on an object, then the object is definitely
  a. at rest.   **c.** being accelerated.
  b. moving with a constant velocity.   **d.** losing mass.

_____ **12.** Which statement about the acceleration of an object is correct?
  a. The acceleration of an object is directly proportional to the net
external force acting on the object and inversely proportional to the
mass of the object.
  b. The acceleration of an object is directly proportional to the net
external force acting on the object and directly proportional to the
mass of the object.
  c. The acceleration of an object is inversely proportional to the net
external force acting on the object and inversely proportional to the
mass of the object.
  d. The acceleration of an object is inversely proportional to the net
external force acting on the object and directly proportional to the
mass of the object.

_____ **13.** Which are simultaneous equal but opposite forces resulting from the
interaction of two objects?
  a. net external forces   **c.** gravitational forces
  b. field forces   **d.** action-reaction pairs

_____ **14.** Newton's third law of motion involves the interactions of
  a. one object and one force.   **c.** two objects and one force.
  b. one object and two forces.   **d.** two objects and two forces.

_____**15.** The magnitude of the gravitational force acting on an object is
   a. frictional force.   **c.** inertia.
   b. weight.   **d.** mass.

_____**16.** A measure of the quantity of matter is
   a. density.   **c.** force.
   b. weight.   **d.** mass.

_____**17.** A change in the gravitational force acting on an object will affect the
   object's
   a. mass.   **c.** weight.
   b. coefficient of static friction.   **d.** inertia.

_____**18.** What are the units of the coefficient of friction?
   a. N   **c.** N^2
   b. 1/N   **d.** The coefficient of friction has
                                                      no units.

SHORT ANSWER

19. In a free-body diagram of an object, why are forces exerted by the object not
included in the diagram?

_____

_____

20. State Newton's first law of motion.

_____

_____

_____

21. In the equation form of Newton's second law, $\Sigma\mathbf{F} = m\mathbf{a}$, what does $\Sigma\mathbf{F}$
represent?

_____

_____

22. What happens to air resistance when an object accelerates?

_____

_____

PROBLEM

23. In a game of tug-of-war, a rope is pulled by a force of 75 N to the left and by a force of 102 N to the right. What is the magnitude and direction of the net horizontal force on the rope?

24. A wagon having a mass of 32 kg is accelerated across a level road at 0.50 m/s². What net force acts on the wagon horizontally?

25. Basking in the sun, a 1.10-kg lizard lies on a flat rock tilted at an angle of 15.0° with respect to the horizontal. What is the magnitude of the normal force exerted by the rock on the lizard?

Assessment

Chapter Test B

Forces and the Laws of Motion
MULTIPLE CHOICE

In the space provided, write the letter of the term or phrase that best completes each statement or best answers each question.

_____ **1.** Which of the following forces is an example of a contact force?
    a. gravitational force   **c.** electric force
    b. magnetic force   **d.** frictional force

_____ **2.** Which of the following forces is an example of a field force?
    a. gravitational force   **c.** normal force
    b. frictional force   **d.** tension

_____ **3.** In the free-body diagram shown to the right, which of the following is the gravitational force acting on the balloon?
    a. 1520 N   **c.** 4050 N
    b. 950 N   **d.** 5120 N

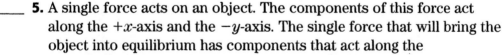

_____ **4.** A late traveler rushes to catch a plane, pulling a suitcase with a force directed 30.0° above the horizontal. If the horizontal component of the force on the suitcase is 60.6 N, what is the force exerted on the handle?
    a. 53.0 N   **c.** 65.2 N
    b. 70.0 N   **d.** 95.6 N

_____ **5.** A single force acts on an object. The components of this force act along the $+x$-axis and the $-y$-axis. The single force that will bring the object into equilibrium has components that act along the
    a. $+x$-axis and $+y$-axis.   **c.** $-x$-axis and $+y$-axis.
    b. $+x$-axis and $-y$-axis.   **d.** $-x$-axis and $-y$-axis.

_____ **6.** As an object falls toward Earth,
    a. the object does not exert a force on Earth.
    b. the object exerts a downward force on Earth.
    c. Newton's third law does not apply.
    d. the upward acceleration of Earth is negligible because of its large mass.

_____ **7.** In general, $\mathbf{F_{net}}$ equals
    a. $\mathbf{F_f}$.   **c.** $\mathbf{F_n}$.
    b. $\mathbf{F_g}$.   **d.** $\Sigma\mathbf{F}$.

Chapter Test B *continued*

_____ **8.** A small force acting on a human-sized object causes
      a. a small acceleration.   **c.** a large acceleration.
      b. no acceleration.   **d.** equilibrium.

_____ **9.** A hammer drives a nail into a piece of wood. Identify an action-reaction pair in this situation.
      a. The nail exerts a force on the hammer; the hammer exerts a force on the wood.
      b. The hammer exerts a force on the nail; the wood exerts a force on the nail.
      c. The hammer exerts a force on the nail; the nail exerts a force on the hammer.
      d. The hammer exerts a force on the nail; the hammer exerts a force on the wood.

_____ **10.** As a basketball player starts to jump for a rebound, the player begins to move upward faster and faster until his shoes leave the floor. At the moment the player begins to jump, the force of the floor on the shoes is
      a. greater than the player's weight.
      b. equal in magnitude and opposite in direction to the player's weight.
      c. less than the player's weight.
      d. zero.

_____ **11.** The magnitude of the gravitational force acting on an object is
      a. frictional force.   **c.** inertia.
      b. weight.   **d.** mass.

_____ **12.** A sled weighing 1.0×10^2 N is held in place on a frictionless 20.0° slope by a rope attached to a stake at the top. The rope is parallel to the slope. What is the normal force of the slope acting on the sled?
      a. 94 N   **c.** 37 N
      b. 47 N   **d.** 34 N

_____ **13.** There are six books in a stack, and each book weighs 5 N. The coefficient of static friction between the books is 0.2. With what horizontal force must one push to start sliding the top five books off the bottom one?
      a. 1 N   **c.** 3 N
      b. 5 N   **d.** 7 N

| Chapter Test B *continued*

SHORT ANSWER

14. Describe how applying the brakes to stop a bicycle is an example of force.

_____

_____

_____

15. Why is force *not* a scalar quantity?

_____

_____

16. Construct a free-body diagram of a car being towed.

17. What is the natural tendency of an object that is in motion?

_____

18. Describe the forces acting on a car as it moves along a level highway in still air at a constant speed.

_____

_____

_____

_____

19. Distinguish between mass and weight.

_____

_____

20. When a car is moving, what happens to the velocity and acceleration of the car if the air resistance becomes equal to the force acting in the opposite direction?

_____

21. Why is air resistance considered a form of friction?

_____

_____

PROBLEM

22. A sled is pulled at a constant velocity across a horizontal snow surface. If a force of 8.0×10^1 N is being applied to the sled rope at an angle of 53° to the ground, what is the magnitude of the force of friction of the snow acting on the sled?

23. A farmhand attaches a 25-kg bale of hay to one end of a rope passing over a frictionless pulley connected to a beam in the hay barn. Another farmhand then pulls down on the opposite end of the rope with a force of 277 N. Ignoring the mass of the rope, what will be the magnitude and direction of the bale's acceleration if the gravitational force acting on it is 245 N?

24. A three-tiered birthday cake rests on a table. From bottom to top, the cake tiers weigh 16 N, 9 N, and 5 N, respectively. What is the magnitude and direction of the normal force acting on the second-tier?

25. A couch with a mass of 1.00×10^2 kg is placed on an adjustable ramp connected to a truck. As one end of the ramp is raised, the couch begins to move downward. If the couch slides down the ramp with an acceleration of 0.70 m/s² when the ramp angle is 25.0°, what is the coefficient of kinetic friction between the ramp and the couch? ($g = 9.81$ m/s²)

Name _____ Class _____ Date _____

Chapter Test A

Work and Energy
MULTIPLE CHOICE

In the space provided, write the letter of the term or phrase that best completes each statement or best answers each question.

_____ **1.** In which of the following sentences is *work* used in the scientific sense of the word?
   a. Holding a heavy box requires a lot of work.
   b. A scientist works on an experiment in the laboratory.
   c. Sam and Rachel pushed hard, but they could do no work on the car.
   d. John learned that shoveling snow is hard work.

_____ **2.** In which of the following sentences is *work* used in the everyday sense of the word?
   a. Lifting a heavy bucket involves doing work on the bucket.
   b. The force of friction usually does negative work.
   c. Sam and Rachel worked hard pushing the car.
   d. Work is a physical quantity.

_____ **3.** A force does work on an object if a component of the force
   a. is perpendicular to the displacement of the object.
   b. is parallel to the displacement of the object.
   c. perpendicular to the displacement of the object moves the object along a path that returns the object to its starting position.
   d. parallel to the displacement of the object moves the object along a path that returns the object to its starting position.

_____ **4.** Work is done when
   a. the displacement is not zero.
   b. the displacement is zero.
   c. the force is zero.
   d. the force and displacement are perpendicular.

_____ **5.** What is the common formula for work?
   a. $W = F\Delta v$ **c.** $W = Fd^2$
   b. $W = Fd$ **d.** $W = F^2d$

_____ **6.** In which of the following scenarios is work done?
   a. A weightlifter holds a barbell overhead for 2.5 s.
   b. A construction worker carries a heavy beam while walking at constant speed along a flat surface.
   c. A car decelerates while traveling on a flat stretch of road.
   d. A student holds a spring in a compressed position.

Chapter Test A *continued*

_____ **7.** In which of the following scenarios is no net work done?
   a. A car accelerates down a hill.
   b. A car travels at constant speed on a flat road.
   c. A car decelerates on a flat road.
   d. A car decelerates as it travels up a hill.

_____ **8.** Which of the following energy forms is associated with an object in motion?
   a. potential energy   **c.** nonmechanical energy
   b. elastic potential energy   **d.** kinetic energy

_____ **9.** Which of the following energy forms is *not* involved in hitting a tennis ball?
   a. kinetic energy   **c.** gravitational potential energy
   b. chemical potential energy   **d.** elastic potential energy

_____ **10.** Which of the following formulas would be used to directly calculate the kinetic energy of a mass bouncing up and down on a spring?
   a. $KE = \frac{1}{2}kx^2$ **c.** $KE = \frac{1}{2}mv^2$

   b. $KE = -\frac{1}{2}kx^2$ **d.** $KE = -\frac{1}{2}mv^2$

_____ **11.** Which of the following equations expresses the work-kinetic energy theorem?
   a. $ME_i = ME_f$ **c.** $\Delta W = KE$
   b. $W_{net} = PE$ **d.** $W_{net} = \Delta KE$

_____ **12.** The main difference between kinetic energy and potential energy is that
   a. kinetic energy involves position, and potential energy involves motion.
   b. kinetic energy involves motion, and potential energy involves position.
   c. although both energies involve motion, only kinetic energy involves position.
   d. although both energies involve position, only potential energy involves motion.

_____ **13.** Which form of energy is involved in weighing fruit on a spring scale?
   a. kinetic energy   **c.** gravitational potential energy
   b. nonmechanical energy   **d.** elastic potential energy

_____ **14.** Gravitational potential energy is always measured in relation to
   a. kinetic energy.   **c.** total potential energy.
   b. mechanical energy.   **d.** a zero level.

_____ **15.** What are the units for a spring constant?
   a. N   **c.** N•m
   b. m   **d.** N/m

| Chapter Test A *continued*

_____**16.** Which of the following is a true statement about the conservation
of energy?
a. Potential energy is always conserved.
b. Kinetic energy is always conserved.
c. Mechanical energy is always conserved.
d. Total energy is always conserved.

_____**17.** Which of the following are examples of conservable quantities?
a. potential energy and length
b. mechanical energy and length
c. mechanical energy and mass
d. kinetic energy and mass

_____**18.** Friction converts kinetic energy to
a. mechanical energy.   **c.** nonmechanical energy.
b. potential energy.   **d.** total energy.

_____**19.** Which of the following is the rate at which work is done?
a. potential energy   **c.** mechanical energy
b. kinetic energy   **d.** power

_____**20.** A more powerful motor can do
a. more work in a longer time interval.
b. the same work in a shorter time interval.
c. less work in a longer time interval.
d. the same work in a longer time interval.

SHORT ANSWER

21. A car travels at a speed of 25 m/s on a flat stretch of road. The driver must
maintain pressure on the accelerator to keep the car moving at this speed.
What is the net work done on the car over a distance of 250 m?

_____

_____

22. State, in words, the work-kinetic energy theorem.

_____

_____

_____

23. A child does 5.0 J of work on a spring while loading a ball into a spring-loaded toy gun. If mechanical energy is conserved, what will be the kinetic energy of the ball when it leaves the gun?

_____

_____

PROBLEM

24. How much work is done on a bookshelf being pulled 5.00 m at an angle of 37.0° from the horizontal? The magnitude of the component of the force that does the work is 43.0 N.

25. What is the average power output of a weightlifter who can lift 250 kg a height of 2.0 m in 2.0 s?

Assessment

Chapter Test B

Work and Energy
MULTIPLE CHOICE

In the space provided, write the letter of the term or phrase that best completes each statement or best answers each question.

_____ **1.** If the sign of work is negative,
  a. the displacement is perpendicular to the force.
  b. the displacement is in the direction opposite the force.
  c. the displacement is in the same direction as the force.
  d. no work is done.

_____ **2.** A child moving at constant velocity carries a 2 N ice-cream cone 1 m across a level surface. What is the net work done on the ice-cream cone?
  a. 0 J   **c.** 2 J
  b. 0.5 J   **d.** 20 J

_____ **3.** A worker does 25 J of work lifting a bucket, then sets the bucket back down in the same place. What is the total net work done on the bucket?
  a. −25 J   **c.** 25 J
  b. 0 J   **d.** 50 J

_____ **4.** If both the mass and the velocity of a ball were tripled, the kinetic energy of the ball would increase by a factor of
  a. 3.   **c.** 9.
  b. 6.   **d.** 27.

_____ **5.** What is the kinetic energy of a 0.135 kg baseball thrown at 40.0 m/s?
  a. 54.0 J   **c.** 108 J
  b. 87.0 J   **d.** 216 J

_____ **6.** If friction is the only force acting on an object during a given physical process, which of the following assumptions can be made in regard to the object's kinetic energy?
  a. The kinetic energy decreases.
  b. The kinetic energy increases.
  c. The kinetic energy remains constant.
  d. The kinetic energy decreases and then increases.

_____ **7.** The equation for determining gravitational potential energy is $PE_g = mgh$. Which factor(s) in this equation is (are) *not* intrinsic to an object?
  a. m **c.** h
  b. g **d.** both g and h

| Chapter Test B *continued*

_____ **8.** Which of the following parameters does *not* depend on how resistant a
spring is to being compressed or stretched?
   a. compression distance   **c.** spring constant
   b. relaxed length   **d.** stretching distance

_____ **9.** What is the potential energy of a 1.0 kg mass 1.0 m above the ground?
   a. 1.0 J   **c.** 10 J
   b. 9.8 J   **d.** 96 J

_____ **10.** Why doesn't the principle of mechanical energy conservation hold in
situations when frictional forces are present?
   a. Kinetic energy is not completely converted to a form of potential
       energy.
   b. Potential energy is completely converted to a form of gravitational
       energy.
   c. Chemical energy is not completely converted to electrical energy.
   d. Kinetic energy is completely converted to a form of gravitational
       energy.

_____ **11.** For which of the following situations is the conservation of
mechanical energy most likely to be a valid assumption?
   a. A skateboard rolls across a sewer grate.
   b. A parachutist falls from a plane.
   c. You rub your hands together to keep warm.
   d. A soccer ball flies through the air.

_____ **12.** A 3.00 kg toy falls from a height of 1.00 m. What will the kinetic energy
of the toy be just before the toy hits the ground? (Assume no air resistance and that $g = 9.81$ m/s^2.)
   a. 98.0 J   **c.** 29.4 J
   b. 0.98 J   **d.** 294 J

_____ **13.** What is the average power supplied by a 60.0 kg person running up a
flight of stairs a vertical distance of 4.0 m in 4.2 s?
   a. 57 W   **c.** 560 W
   b. 240 W   **d.** 670 W

SHORT ANSWER

14. Explain the scientific meaning of *work*.

_____

_____

_____

_____

| **Chapter Test B** *continued*

15. What form of energy is associated with the position of an object in Earth's gravitational field?

_____

16. Describe the relationship between kinetic energy and gravitational potential energy during the free fall of a pencil from a desk.

_____

_____

_____

_____

17. Write an equation that expresses the conservation of mechanical energy in a system that involves kinetic energy, gravitational potential energy, and elastic potential energy.

_____

18. Show how the alternative definition of *power* can be derived by substituting the definitions of *work* and *speed* into the standard definition of *power*,

$$P = \frac{W}{\Delta t}$$

19. Which motor performs more work in the same amount of time—a 10 kW motor or a 20 kW motor? How much more work can it do?

_____

_____

_____

PROBLEM

20. What is the kinetic energy of a 1.5×10^3 kg car traveling at 25 m/s?

21. A worker pushes a box with a horizontal force of 50.0 N over a level distance of 5.0 m. If a frictional force of 43 N acts on the box in a direction opposite to that of the worker, what net work is done on the box ?

22. A professional skier starts from rest and reaches a speed of 56 m/s on a ski slope angled 30.0° above the horizontal. Using the work-kinetic energy theorem and disregarding friction, find the minimum distance along the slope the skier would have to travel in order to reach this speed.

23. An 80.0 kg climber climbs to the top of Mount Everest, which has a peak height of 8848 m. What is the climber's potential energy with respect to sea level?

24. Old Faithful geyser in Yellowstone National Park shoots water every hour to a height of 40.0 m. With what velocity does the water leave the ground? (Assume no air resistance and that $g = 9.81$ m/s^2.)

25. Water flows over a section of Niagara Falls at a rate of 1.20×10^6 kg/s and falls 50.0 m. What is the power of the waterfall?

Chapter Test A

Momentum and Collisions
MULTIPLE CHOICE

In the space provided, write the letter of the term or phrase that best completes each statement or best answers each question.

_____ **1.** When comparing the momentum of two moving objects, which of the following is correct?
   a. The object with the higher velocity will have less momentum if the masses are equal.
   b. The more massive object will have less momentum if its velocity is greater.
   c. The less massive object will have less momentum if the velocities are the same.
   d. The more massive object will have less momentum if the velocities are the same.

_____ **2.** A child with a mass of 23 kg rides a bike with a mass of 5.5 kg at a velocity of 4.5 m/s to the south. Compare the momentum of the child with the momentum of the bike.
   a. Both the child and the bike have the same momentum.
   b. The bike has a greater momentum than the child.
   c. The child has a greater momentum than the bike.
   d. Neither the child nor the bike has momentum.

_____ **3.** A roller coaster climbs up a hill at 4 m/s and then zips down the hill at 30 m/s. The momentum of the roller coaster
   a. is greater up the hill than down the hill.
   b. is greater down the hill than up the hill.
   c. remains the same throughout the ride.
   d. is zero throughout the ride.

_____ **4.** If a force is exerted on an object, which statement is true?
   a. A large force always produces a large change in the object's momentum.
   b. A large force produces a large change in the object's momentum only if the force is applied over a very short time interval.
   c. A small force applied over a long time interval can produce a large change in the object's momentum.
   d. A small force produces a large change in the object's momentum.

| Chapter Test A *continued*

_____ **5.** A ball with a momentum of 4.0 kg•m/s hits a wall and bounces straight back without losing any kinetic energy. What is the change in the ball's momentum?
  a. −8.0 kg•m/s
  b. −4.0 kg•m/s
  c. 0.0 kg•m/s
  d. 8.0 kg•m/s

_____ **6.** The impulse experienced by a body is equivalent to the body's change in
  a. velocity.
  b. kinetic energy.
  c. momentum.
  d. force.

_____ **7.** A 75 kg person walking around a corner bumped into an 80 kg person who was running around the same corner. The momentum of the 80 kg person
  a. increased.
  b. decreased.
  c. remained the same.
  d. was conserved.

_____ **8.** Two skaters stand facing each other. One skater's mass is 60 kg, and the other's mass is 72 kg. If the skaters push away from each other without spinning,
  a. the lighter skater has less momentum.
  b. their momenta are equal but opposite.
  c. their total momentum doubles.
  d. their total momentum decreases.

_____ **9.** In a two-body collision,
  a. momentum is always conserved.
  b. kinetic energy is always conserved.
  c. neither momentum nor kinetic energy is conserved.
  d. both momentum and kinetic energy are always conserved.

_____**10.** The law of conservation of momentum states that
  a. the total initial momentum of all objects interacting with one another usually equals the total final momentum.
  b. the total initial momentum of all objects interacting with one another does not equal the total final momentum.
  c. the total momentum of all objects interacting with one another is zero.
  d. the total momentum of all objects interacting with one another remains constant regardless of the nature of the forces between the objects.

_____**11.** Two objects stick together and move with a common velocity after colliding. Identify the type of collision.
  a. elastic
  b. perfectly elastic
  c. inelastic
  d. perfectly inelastic

_____**12.** Two billiard balls collide. Identify the type of collision.
  a. elastic
  b. perfectly elastic
  c. inelastic
  d. perfectly inelastic

_____ **13.** In an inelastic collision between two objects with unequal masses,
   a. the total momentum of the system will increase.
   b. the total momentum of the system will decrease.
   c. the kinetic energy of one object will increase by the amount that the kinetic energy of the other object decreases.
   d. the momentum of one object will increase by the amount that the momentum of the other object decreases.

_____ **14.** A billiard ball collides with a stationary identical billiard ball in an elastic head-on collision. After the collision, which of the following is true of the first ball?
   a. It maintains its initial velocity.
   b. It has one-half its initial velocity.
   c. It comes to rest.
   d. It moves in the opposite direction.

SHORT ANSWER

15. As a bullet travels through the air, it slows down due to air resistance. How does the bullet's momentum change as a result?

_____

_____

16. A student walks to class at a velocity of 3 m/s. To avoid walking into a door as it opens, the student slows to a velocity of 0.5 m/s. Now late for class, the student runs down the corridor at a velocity of 7 m/s. At what point in this scenario does the student have the least momentum?

_____

_____

17. How can a small force produce a large change in momentum?

_____

_____

18. Two billiard balls of equal mass are traveling straight toward each other with the same speed. They meet head-on in an elastic collision. What is the total momentum of the system containing the two balls before the collision?

_____

| Chapter Test A *continued*

PROBLEM

19. Compare the momentum of a 6160 kg truck moving at 3.00 m/s to the momentum of a 1540 kg car moving at 12.0 m/s.

20. A ball with a mass of 0.15 kg and a velocity of 5.0 m/s strikes a wall and bounces straight back with a velocity of 3.0 m/s. What is the change in momentum of the ball?

Name _____ Class _____ Date _____

Chapter Test B

Momentum and Collisions
MULTIPLE CHOICE

In the space provided, write the letter of the term or phrase that best completes each statement or best answers each question.

_____ **1.** Which of the following has the greatest momentum?
    a. a tortoise with a mass of 275 kg moving at a velocity of 0.55 m/s
    b. a hare with a mass of 2.7 kg moving at a velocity of 7.5 m/s
    c. a turtle with a mass of 91 kg moving at a velocity of 1.4 m/s
    d. a roadrunner with a mass of 1.8 kg moving at a velocity of 6.7 m/s

_____ **2.** A person sitting in a chair with wheels stands up, causing the chair to roll backward across the floor. The momentum of the chair
    a. was zero while stationary and increased when the person stood.
    b. was greatest while the person sat in the chair.
    c. remained the same.
    d. was zero when the person got out of the chair and increased while the person sat.

_____ **3.** A 0.2 kg baseball is pitched with a velocity of 40 m/s and is then batted to the pitcher with a velocity of 60 m/s. What is the magnitude of change in the ball's momentum?
    a. 2 kg•m/s   **c.** 8 kg•m/s
    b. 4 kg•m/s   **d.** 20 kg•m/s

_____ **4.** Which of the following statements properly relates the variables in the equation $\mathbf{F}\Delta t = \Delta\mathbf{p}$?
    a. A large constant force changes an object's momentum over a long time interval.
    b. A large constant force acting over a long time interval causes a large change in momentum.
    c. A large constant force changes an object's momentum at various time intervals.
    d. A large constant force does not necessarily cause a change in an object's momentum.

_____ **5.** Two objects with different masses collide and bounce back after an elastic collision. Before the collision, the two objects were moving at velocities equal in magnitude but opposite in direction. After the collision,
    a. the less massive object had gained momentum.
    b. the more massive object had gained momentum.
    c. both objects had the same momentum.
    d. both objects lost momentum.

_____ **6.** Two swimmers relax close together on air mattresses in a pool. One swimmer's mass is 48 kg, and the other's mass is 55 kg. If the swimmers push away from each other,

a. their total momentum triples.

b. their momenta are equal but opposite.

c. their total momentum doubles.

d. their total momentum decreases.

_____ **7.** Which of the following statements about the conservation of momentum is *not* correct?

a. Momentum is conserved for a system of objects pushing away from each other.

b. Momentum is not conserved for a system of objects in a head-on collision.

c. Momentum is conserved when two or more interacting objects push away from each other.

d. The total momentum of a system of interacting objects remains constant regardless of forces between the objects.

_____ **8.** Two balls of dough collide and stick together. Identify the type of collision.

a. elastic   **c.** inelastic

b. perfectly elastic   **d.** perfectly inelastic

_____ **9.** Which of the following best describes the momentum of two bodies after a two-body collision if the kinetic energy of the system is conserved?

a. must be less   **c.** might also be conserved

b. must also be conserved   **d.** is doubled in value

SHORT ANSWER

10. A baseball pitcher's first pitch is a fastball, moving at high speed. The pitcher's second pitch—with the same ball—is a changeup, moving more slowly. Which pitch is harder for the catcher to stop? Explain your answer in terms of momentum.

_____

_____   _____

11. Is it possible for a spaceship traveling with constant velocity to experience a change in momentum? Explain your answer.

_____

_____

12. Each croquet ball in a set has a mass of 0.50 kg. The green ball travels at 10.5 m/s and strikes a stationary red ball. If the green ball stops moving, what is the final speed of the red ball after the collision?

_____

13. A moderate force will break an egg. Using the concepts of momentum, force, and time interval, explain why an egg is more likely to break when it is dropped on concrete than if it is dropped on grass.

_____

_____

_____

14. Why is the sound produced by a collision evidence that the collision is not perfectly elastic?

_____

_____

_____

PROBLEM

15. What velocity must a 1340 kg car have in order to have the same momentum as a 2680 kg truck traveling at a velocity of 15 m/s to the west?

16. A 6.0×10^{-2} kg tennis ball moves at a velocity of 12 m/s. The ball is struck by a racket, causing it to rebound in the opposite direction at a speed of 18 m/s. What is the change in the ball's momentum?

17. A train with a mass of 1.8×10^3 kg is moving at 15 m/s when the engineer applies the brakes. If the braking force is constant at 3.5×10^4 N, how long does it take the train to stop? How far does the train travel during this time?

18. An astronaut with a mass of 85 kg is outside a space capsule when the tether line breaks. To return to the capsule, the astronaut throws a 2.0 kg wrench away from the capsule at a speed of 14 m/s. At what speed does the astronaut move toward the capsule?

19. A 0.10 kg object makes an elastic head-on collision with a 0.15 kg stationary object. The final velocity of the 0.10 kg object after the collision is -0.045 m/s, and the final velocity of the 0.15 kg object after the collision is 0.16 m/s. What was the initial velocity of the 0.10 kg object?

20. A 90 kg halfback runs north and is tackled by a 120 kg opponent running south at 4 m/s. The collision is perfectly inelastic. Just after the tackle, both players move at a velocity of 2 m/s north. Calculate the velocity of the 90 kg player just before the tackle.

Chapter Test A

Circular Motion and Gravitation

MULTIPLE CHOICE

In the space provided, write the letter of the term or phrase that best completes each statement or best answers each question.

_____ **1.** When an object is moving with uniform circular motion, the object's tangential speed
   a. is circular.
   b. is perpendicular to the plane of motion.
   c. is constant.
   d. is directed toward the center of motion.

_____ **2.** The centripetal force on an object in circular motion is
   a. in the same direction as the tangential speed.
   b. in the direction opposite the tangential speed.
   c. in the same direction as the centripetal acceleration.
   d. in the direction opposite the centripetal acceleration.

_____ **3.** A ball is whirled on a string, then the string breaks. What causes the ball to move off in a straight line?
   a. centripetal acceleration   **c.** centrifugal force
   b. centripetal force   **d.** inertia

_____ **4.** If you lift an apple from the ground to some point above the ground, the gravitational potential energy in the system increases. This potential energy is stored in
   a. the apple.
   b. Earth.
   c. both the apple and Earth.
   d. the gravitational field between Earth and the apple.

_____ **5.** When calculating the gravitational force between two extended bodies, you should measure the distance
   a. from the closest points on each body.
   b. from the most distant points on each body.
   c. from the center of each body.
   d. from the center of one body to the closest point on the other body.

_____ **6.** The gravitational force between two masses is 36 N. What is the gravitational force if the distance between them is tripled?
   ($G = 6.673 \times 10^{-11}$ N•m^2/kg^2)
   a. 4.0 N   **c.** 18 N
   b. 9.0 N   **d.** 27 N

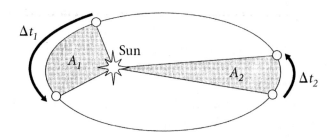

_____ **7.** In the figure above, according to Kepler's laws of planetary motion,
    a. $A_1 = A_2$.
    b. $\Delta t_1 > \Delta t_2$.
    c. if $\Delta t_1 = \Delta t_2$, then the orbit is circular.
    d. if $\Delta t_1 = \Delta t_2$, then $A_1 = A_2$.

_____ **8.** Newton's law of universal gravitation
    a. can be derived from Kepler's laws of planetary motion.
    b. can be used to derive Kepler's third law of planetary motion.
    c. can be used to disprove Kepler's laws of planetary motion.
    d. does not apply to Kepler's laws of planetary motion.

_____ **9.** The equation for the speed of an object in circular orbit is $v_t = \sqrt{G\dfrac{m}{r}}$.
    What does m represent in this equation?
    a. the mass of the sun
    b. the mass of Earth
    c. the mass of the central object
    d. the mass of the orbiting object

_____ **10.** How would the speed of Earth's orbit around the sun change if Earth's mass increased by 4 times?
    a. It would increase by a factor of 2.
    b. It would increase by a factor of 4.
    c. It would decrease by a factor of 2.
    d. The speed would not change.

_____ **11.** When an astronaut in orbit experiences apparent weightlessness,
    a. no forces act on the astronaut.
    b. no gravitational forces act on the astronaut.
    c. the net gravitational force on the astronaut is zero.
    d. the net gravitational force on the astronaut is not balanced by a normal force.

_____ **12.** If you cannot exert enough force to loosen a bolt with a wrench, which of the following should you do?
    a. Use a wrench with a longer handle.
    b. Tie a rope to the end of the wrench and pull on the rope.
    c. Use a wrench with a shorter handle.
    d. You should exert a force on the wrench closer to the bolt.

_____**13.** Suppose a doorknob is placed at the center of a door. Compared with a door whose knob is located at the edge, what amount of force must be applied to this door to produce the torque exerted on the other door?
  a. one-half as much   **c.** one-fourth as much
  b. two times as much   **d.** four times as much

_____**14.** A heavy bank-vault door is opened by the application of a force of 3.0×10^2 N directed perpendicular to the plane of the door at a distance of 0.80 m from the hinges. What is the torque?
  a. 120 N•m   **c.** 300 N•m
  b. 240 N•m   **d.** 360 N•m

_____**15.** What kind of simple machine are you using if you pry a nail from a board with the back of a hammer?
  a. a wedge   **c.** a lever
  b. a pulley   **d.** a screw

_____**16.** Which of the following is *not* a valid equation for mechanical advantage?
  a. $MA = \dfrac{F_{out}}{F_{in}}$ **c.** $MA = \dfrac{W_{out}}{W_{in}}$

  b. $MA = \dfrac{d_{in}}{d_{out}}$ **d.** $MA = \dfrac{\text{output force}}{\text{input force}}$

SHORT ANSWER

17. Explain how an object moving at a constant speed can have a nonzero acceleration.

_____

_____

_____

18. What provides the centripetal force for a car driving on a circular track?

_____

_____

_____

19. Is there an outward force in circular motion? Explain.

_____

_____

_____

Chapter Test A *continued*

20. Discuss the following statement: "A satellite is continually in free fall."

_____

_____

_____

21. Earth exerts a 1.0 N gravitational force on an apple. Does the apple accelerate toward Earth, or does Earth accelerate toward the apple? Explain your answer.

_____

_____

_____

22. Are astronauts in orbit weightless? Explain your answer.

_____

_____

_____

PROBLEM

23. A 35 kg child moves with uniform circular motion while riding a horse on a carousel. The horse is 3.2 m from the carousel's axis of rotation and has a tangential speed of 2.6 m/s. What is the child's centripetal acceleration?

24. What is the centripetal force on the child in item 23?

25. A force of 255 N is needed to pull a nail from a wall, using a claw hammer. If the resistance force of the nail is 3650 N, what is the mechanical advantage of the hammer?

Chapter Test B

Circular Motion and Gravitation
MULTIPLE CHOICE

In the space provided, write the letter of the term or phrase that best completes each statement or best answers each question.

_____ **1.** What term describes a change in the speed of an object in circular motion?
- **a.** tangential speed
- **b.** tangential acceleration
- **c.** centripetal acceleration
- **d.** centripetal force

_____ **2.** When a car makes a sharp left turn, what causes the passengers to move toward the right side of the car?
- **a.** centripetal acceleration
- **b.** centripetal force
- **c.** centrifugal force
- **d.** inertia

_____ **3.** Two small masses that are 10.0 cm apart attract each other with a force of 10.0 N. When they are 5.0 cm apart, these masses will attract each other with what force? ($G = 6.673 \times 10^{-11}$ N•m^2/kg^2)
- **a.** 5.0 N
- **b.** 2.5 N
- **c.** 20.0 N
- **d.** 40.0 N

_____ **4.** How would the speed of Earth's orbit around the sun change if Earth's distance from the sun increased by 4 times?
- **a.** It would increase by a factor of 2.
- **b.** It would increase by a factor of 4.
- **c.** It would decrease by a factor of 2.
- **d.** The speed would not change.

_____ **5.** Which of the following statements is correct?
- **a.** The farther the force is from the axis of rotation, the more torque is produced.
- **b.** The closer the force is to the axis of rotation, the more torque is produced.
- **c.** The closer the force is to the axis of rotation, the easier it is to rotate the object.
- **d.** The farther the force is from the axis of rotation, the less torque is produced.

_____ **6.** If the torque required to loosen a nut on a wheel has a magnitude of 40.0 N•m and the force exerted by a mechanic is 133 N, how far from the nut must the mechanic apply the force?
- **a.** 1.20 m
- **b.** 15.0 cm
- **c.** 30.1 cm
- **d.** 60.2 cm

_____ **7.** An iron bar is used to lift a slab of cement. The force applied to lift the slab is 4.0×10^2 N. If the slab weighs 6400 N, what is the mechanical advantage of the bar?

   a. 1.6   **c.** 6000

   b. 16   **d.** 6.3%

SHORT ANSWER

8. Two horses are side by side on a carousel. Which has a greater tangential speed—the one closer to the center or the one farther from the center? Explain your answer.

_____

_____

_____

9. Show how the equation for centripetal force can be derived by substituting the equation for centripetal acceleration into Newton's second law.

10. Earth exerts a 1.0 N gravitational force on an apple. What is the magnitude of the gravitational force the apple exerts on Earth?

_____

11. What law does the diagram shown above illustrate?

_____

_____

| Chapter Test B *continued*

12. How did Newton use Kepler's laws?

_____

_____

_____

_____

13. How will an object move if it is acted on by a positive net torque and a net force of zero?

_____

_____

14. Explain how the operation of a simple machine alters the applied force and the distance moved.

_____

_____

PROBLEM

15. A car on a roller coaster loaded with passengers has a mass of 2.0×10^3 kg. At the lowest point of the track, the radius of curvature of the track is 24 m and the roller car has a tangential speed of 17 m/s. What is the centripetal acceleration of the roller car at the lowest point on the track?

16. What force is exerted on the roller car in item 15 by the track at the lowest point?

17. Two trucks with equal mass are attracted to each other with a gravitational force of 6.7×10^{-4} N. The trucks are separated by a distance of 3.0 m. What is the mass of one of the trucks? ($G = 6.673 \times 10^{-11}$ N•m^2/kg^2)

18. Show how the constant of proportionality in Kepler's third law can be found by equating gravitational force and the centripetal force in a circular orbit.

19. A new moon is discovered orbiting Neptune with an orbital speed of 9.3×10^3 m/s. Neptune's mass is 1.0×10^{26} kg. What is the radius of the new moon's orbit? What is the orbital period? Assume that the orbit is circular. ($G = 6.673 \times 10^{-11}$ N•m^2/kg^2)

20. A force of 4.0 N is applied to a door at an angle of 60.0° and a distance of 0.30 m from the hinge. What is the torque produced?

Chapter Test A

Fluid Mechanics
MULTIPLE CHOICE

In the space provided, write the letter of the term or phrase that best completes each statement or best answers each question.

_____ **1.** Which of the following is a fluid?
  a. helium
  b. ice
  c. iron
  d. gold

_____ **2.** Which of the following statements is *not* correct?
  a. A fluid flows.
  b. A fluid has a definite shape.
  c. Molecules of a fluid are free to move past each other.
  d. A fluid changes its shape easily.

_____ **3.** How does a liquid differ from a gas?
  a. A liquid has both definite shape and definite volume, whereas a gas has neither.
  b. A liquid has definite volume, unlike a gas.
  c. A liquid has definite shape, unlike a gas.
  d. A liquid has definite shape, whereas a gas has definite volume.

_____ **4.** For incompressible fluids, density changes little with changes in
  a. depth.
  b. temperature.
  c. pressure.
  d. free-fall acceleration.

_____ **5.** A cube of wood with a density of 0.780 g/cm^3 is 10.0 cm on each side. When the cube is placed in water, what buoyant force acts on the wood? ($\rho_w = 1.00$ g/cm^3)
  a. 7.65×10^3 N
  b. 7.65 N
  c. 6.40 N
  d. 5.00 N

_____ **6.** A buoyant force acts in the opposite direction of gravity. Therefore, which of the following is true of an object completely submerged in water?
  a. The net force on the object is smaller than the weight of the object.
  b. The net force on the object is larger than the weight of the object.
  c. The net force on the object is equal to the weight of the object.
  d. The object appears to weigh more than it does in air.

_____ **7.** Which of the following statements about floating objects is correct?
  a. The object's density is greater than the density of the fluid on which it floats.
  b. The object's density is equal to the density of the fluid on which it floats.
  c. The displaced volume of fluid is greater than the volume of the object.
  d. The buoyant force equals the object's weight.

_____ **8.** Which of the following statements is true according to Pascal's principle?
  a. Pressure in a fluid is greatest at the walls of the container holding the fluid.
  b. Pressure in a fluid is greatest at the center of the fluid.
  c. Pressure in a fluid is the same throughout the fluid.
  d. Pressure in a fluid is greatest at the top of the fluid.

_____ **9.** A water bed that is 1.5 m wide and 2.5 m long weighs 1055 N. Assuming the entire lower surface of the bed is in contact with the floor, what is the pressure the bed exerts on the floor?
  a. 250 Pa   **c.** 270 Pa
  b. 260 Pa   **d.** 280 Pa

_____ **10.** What factors affect the gauge pressure within a fluid?
  a. fluid density, depth, free-fall acceleration
  b. fluid volume, depth, free-fall acceleration
  c. fluid mass, depth, free-fall acceleration
  d. fluid weight, depth, free-fall acceleration

_____ **11.** If the air pressure in a tire is measured as 2.0×10^5 Pa, and atmospheric pressure equals 1.0×10^5 Pa, what pressure does the air within the tire exert outward on the tire walls?
  a. 1.0×10^5 Pa   **c.** 3.0×10^5 Pa
  b. 2.0×10^5 Pa   **d.** 4.0×10^5 Pa

_____ **12.** Which of the following properties is *not* characteristic of an ideal fluid?
  a. laminar flow   **c.** nonviscous
  b. turbulent flow   **d.** incompressible

_____ **13.** Which of the following is *not* an example of laminar flow?
  a. a river moving slowly in a straight line
  b. smoke rising upward in a smooth column through air
  c. water flowing evenly from a slightly opened faucet
  d. smoke twisting as it moves upward from a fire

_____**14.** Which of the following is *not* an example of turbulent flow?
  a. a river flowing slowly in a straight line
  b. a river flowing swiftly around rocks in rapids
  c. water flowing unevenly from a fully opened faucet
  d. smoke twisting as it moves upward from a fire

_____**15.** Why does an ideal fluid move faster through a pipe with decreasing diameter?
  a. The pressure within the fluid increases.
  b. The pressure within the fluid decreases.
  c. The pipe exerts more pressure on the fluid.
  d. The fluid moves downhill.

_____**16.** Why does the lift on an airplane wing increase as the speed of the airplane increases?
  a. The pressure behind the wing becomes less than the pressure in front of the wing.
  b. The pressure behind the wing becomes greater than the pressure in front of the wing.
  c. The pressure above the wing becomes less than the pressure below the wing.
  d. The pressure above the wing becomes greater than the pressure below the wing.

SHORT ANSWER

17. Why are solid objects not considered to be fluids?

_____

_____

18. How does a gas change shape when it is poured from a small flask into a large flask?

_____

_____

19. Why is the net force on a submerged object called its *apparent weight*?

_____

_____

_____

20. What determines whether an object will sink or float?

_____

_____

21. Describe how Pascal's principle allows the pressure throughout a fluid to be known.

_____

_____

22. What does Bernoulli's principle state will happen to the pressure in a fluid as the speed of the fluid increases?

_____

23. Use Bernoulli's principle to explain why a nozzle on a fire hose is tapered.

_____

_____

_____

PROBLEM

24. An ice cube is placed in a glass of water. The cube is 2.0 cm on each side and has a density of 0.917 g/cm^3. What is the magnitude of the buoyant force on the ice?

25. A hydraulic lift consists of two pistons that connect to each other by an incompressible fluid. If one piston has an area of 0.15 m^2 and the other an area of 6.0 m^2, how large a mass can be raised by a force of 130 N exerted on the smaller piston?

Assessment

Chapter Test B

Fluid Mechanics
MULTIPLE CHOICE

In the space provided, write the letter of the term or phrase that best completes each statement or best answers each question.

_____ **1.** Which of the following is *not* a fluid?
   a. carbon dioxide   **c.** seawater
   b. hydrogen   **d.** wood

_____ **2.** When a gas is poured out of one container into another container, which of the following does *not* occur?
   a. The gas flows into the new container.
   b. The gas changes shape to fit the new container.
   c. The gas spreads out to fill the new container.
   d. The gas keeps its original volume.

_____ **3.** According to legend, to determine whether the king's crown was made of pure gold, Archimedes measured the crown's volume by determining how much water it displaced. The density of gold is 19.3 g/cm^3. If the crown's mass was 6.00×10^2 g, what volume of water would have been displaced if the crown was indeed made of pure gold?
   a. 31.1 cm^3 **c.** 22.8×10^3 cm^3
   b. 1.81×10^3 cm^3 **d.** 114×10^3 cm^3

_____ **4.** Which of the following statements about completely submerged objects resting on the ocean bottom is correct?
   a. The buoyant force acting on the object is equal to the object's weight.
   b. The apparent weight of the object depends on the object's density.
   c. The displaced volume of fluid is greater than the volume of the object.
   d. The weight of the object and the buoyant force are equal and opposite.

_____ **5.** If an object is only partially submerged in a fluid, which of the following is true?
   a. The volume of the displaced fluid equals the volume of the object.
   b. The density of the fluid equals the density of the object.
   c. The density of the fluid is greater than the density of the object.
   d. The density of the fluid is less than the density of the object.

_____ **6.** Each of four tires on an automobile has an area of 0.026 m^2 in contact with the ground. The weight of the automobile is 2.6×10^4 N. What is the pressure in the tires?
   a. 3.1×10^6 Pa   **c.** 2.5×10^5 Pa
   b. 1.0×10^6 Pa   **d.** 6.5×10^3 Pa

_____ **7.** A force of 230 N applied on a hydraulic lift raises an automobile weighing 6500 N. If the applied force is exerted on a 7.0 m² piston, what is the area of the piston beneath the automobile?
   a. 2.0×10^2 m²   **c.** 0.25 m²
   b. 0.0050 m²   **d.** 4.0 m²

_____ **8.** The gauge pressure for the air in a balloon equals 1.01×10^5 Pa. If atmospheric pressure is equal to 1.01×10^5 Pa, what is the absolute pressure of the air inside the balloon?
   a. 0 Pa   **c.** 1.01×10^5 Pa
   b. 5.05×10^4 Pa   **d.** 2.02×10^5 Pa

_____ **9.** A closed vessel can sink to a depth of 20.0 m in water ($\rho_w = 1.00$ g/cm³) before the external pressure crushes it. To what depth could this same container be immersed in a deep vat of mercury ($\rho_{Hg} = 13.6$ g/cm³) without it being crushed?
   a. 0.680 m   **c.** 15.7 m
   b. 1.47 m   **d.** 27.2 m

_____ **10.** An ideal fluid flows through a pipe made of two sections with diameters of 1 cm and 3 cm, respectively. By what factor would you have to multiply the velocity of the liquid flowing through the 1 cm section to obtain the velocity of liquid flowing through the 3 cm section?

   a. 9   **c.** $\dfrac{1}{3}$

   b. 6   **d.** $\dfrac{1}{9}$

SHORT ANSWER

11. A passenger in a descending balloon releases sand from a bag attached to the balloon's gondola. This causes the balloon to stop moving downward and to remain at a constant elevation. Explain both sets of motion for the balloon in terms of Archimedes' principle.

_____

_____

_____

12. Four liquids with densities of 1.260 g/cm³, 0.690 g/cm³, 0.970 g/cm³, and 0.870 g/cm³ are poured into a container. The liquids form separate layers, with one liquid floating above another. What are the densities of the liquids, in order from the liquid at the top to the one at the bottom?

_____

_____

13. The piston diameters for a hydraulic lift have a ratio of 4 to 1. What must the minimum ratio of the force on the smaller piston to that on the larger piston be for the lift to operate?

_____

14. Why is the base of a dam thicker than the top part of the dam?

_____

_____

_____

15. An aspirator uses the laminar flow of water through a tube to pull air outside the tube into the tube. Use Bernoulli's principle to explain how an aspirator works.

_____

_____

_____

PROBLEM

16. A ball with a density of 0.940 g/cm^3 and a volume of 1.4×10^4 cm^3 is placed in a fluid with a density of 0.870 g/cm^3. Does the ball sink or float? If the ball floats, calculate the volume of the displaced fluid. If the ball sinks, calculate the magnitude of the apparent weight of the ball.

17. A piece of wood with a mass of 6.88 kg is placed in fresh water ($\rho_w = 1.00$ g/cm^3). What is the density of the wood if it has an apparent weight of −6.13 N?

18. The absolute pressure below the surface of a freshwater lake is 3.51×10^5 Pa. At what depth does this pressure occur? Assume that atmospheric pressure is 1.01×10^5 Pa and the density of the water is 1.00×10^3 kg/m^3.

19. A circular hatch in the hull of a submarine has a radius of 40.0 cm. The submarine is 850.0 m under water. If atmospheric pressure above the ocean is 1.01×10^5 Pa and the air pressure inside the submarine is 1.31×10^5 Pa, what net force is exerted on the hatch? ($\rho_{sw} = 1025$ kg/m^3)

20. Water flows at a speed of 15 m/s through a pipe that has a radius of 0.40 m. The water then flows through a smaller pipe at a speed of 45 m/s. What is the radius of the smaller pipe?

Chapter Test A

Heat

MULTIPLE CHOICE

In the space provided, write the letter of the term or phrase that best completes each statement or best answers each question.

_____ 1. Which of the following is proportional to the kinetic energy of atoms and molecules?
    a. elastic energy   **c.** potential energy
    b. temperature   **d.** thermal equilibrium

_____ 2. Which of the following is a form of kinetic energy that occurs within a molecule when the bonds are stretched or bent?
    a. translational   **c.** vibrational
    b. rotational   **d.** internal

_____ 3. As the temperature of a substance increases, its volume tends to increase due to
    a. thermal equilibrium.   **c.** thermal expansion.
    b. thermal energy.   **d.** thermal contraction.

_____ 4. If two small beakers of water, one at 70°C and one at 80°C, are emptied into a large beaker, what is the final temperature of the water?
    a. The final temperature is less than 70°C.
    b. The final temperature is greater than 80°C.
    c. The final temperature is between 70°C and 80°C.
    d. The water temperature will fluctuate.

_____ 5. A substance registers a temperature change from 20°C to 40°C. To what incremental temperature change does this correspond?
    a. 20 K   **c.** 36 K
    b. 40 K   **d.** 313 K

_____ 6. Energy transferred as heat occurs between two bodies in thermal contact when they differ in which of the following properties?
    a. mass   **c.** density
    b. specific heat   **d.** temperature

_____ 7. The use of fiberglass insulation in the outer walls of a building is intended to minimize heat transfer through what process?
    a. conduction   **c.** convection
    b. radiation   **d.** vaporization

_____ **8.** Energy transfer as heat between two objects depends on which of the following?

    a. The difference in mass of the two objects.

    b. The difference in volume of the two objects.

    c. The difference in temperature of the two objects.

    d. The difference in composition of the two objects.

_____ **9.** Why does sandpaper get hot when it is rubbed against rusty metal?

    a. Energy is transferred from the sandpaper into the metal.

    b. Energy is transferred from the metal to the sandpaper

    c. Friction between the sandpaper and metal increases the temperature of both.

    d. Energy is transferred from a hand to the sandpaper.

_____ **10.** In the presence of friction, not all of the work done on a system appears as mechanical energy. What happens to the rest of the energy provided by work?

    a. The remaining energy is stored as mechanical energy within the system.

    b. The remaining energy is dissipated as sound.

    c. The remaining energy causes a decrease in the internal energy of the system.

    d. The remaining energy causes an increase in the internal energy of the system.

_____ **11.** A nail is driven into a board with an initial kinetic energy of 150 J. If the potential energy before and after the event is the same, what is the change in the internal energy of the board and nail?

    a. 150 J   **c.** 0 J

    b. 75 J   **d.** −150 J

_____ **12.** A calorimeter is used to determine the specific heat capacity of a test metal. If the specific heat capacity of water is known, what quantities must be measured?

    a. metal volume, water volume, initial and final temperatures of metal and water

    b. metal mass, water mass, initial and final temperatures of metal and water

    c. metal mass, water mass, final temperature of metal and water

    d. metal mass, water mass, heat added to or removed from water and metal

_____ **13.** Which of the following describes a substance in which the temperature and pressure remain constant while the substance experiences an inward transfer of energy?

    a. gas   **c.** solid

    b. liquid   **d.** substance undergoing a change of state

_____**14.** Which of the following is true during a phase change?
   a. Temperature increases.
   b. Temperature remains constant.
   c. Temperature decreases.
   d. There is no transfer of energy as heat.

SHORT ANSWER

15. Describe how temperature is related to the kinetic energy of the particles of the gas in the figure above.

_____

_____

16. A pan of water at a temperature of 80°C is placed on a block of porcelain at a temperature of 15°C. What can you state about the temperatures of the objects when they are in thermal equilibrium?

_____

_____

17. If energy is transferred as heat from a closed metal container to the air surrounding it, what is true of the initial temperatures of each?

_____

_____

18. Two bottles are immersed in tubs of water. In one case, the bottle's temperature is 40°C, and the water's temperature is 20°C. In the other case, the bottle's temperature is 55°C, and the water's temperature is 35°C. How does the energy transfer between the bottles and the water differ for the two cases?

_____

_____

19. What is the specific heat capacity of a substance?

_____

_____

| Chapter Test A *continued*

The figure below shows how the temperature of 10.0 g of ice changes as energy is added. Use the figure to answer questions 20–22.

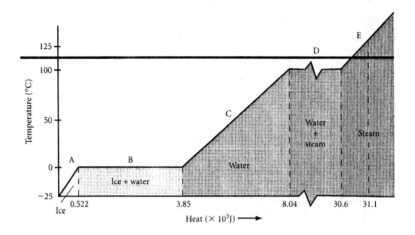

20. What happens to the ice at 0°C?

_____

21. What happens to the ice at 100°C?

_____

22. What happens to the ice between 0°C and 100°C?

_____

_____

PROBLEM

23. A warm day has a high temperature of 38.0°C. What is this temperature in degrees Fahrenheit?

24. What is the increase in the internal energy per kilogram of water at the bottom of a 145 m waterfall, assuming that all of the initial potential energy is transferred as heat to the water? ($g = 9.81 \ m/s^2$)

25. What is the temperature increase of 4.0 kg of water when it is heated by an 8.0×10^2 W immersion heater for exactly 10.0 min? ($c_p = 4186$ J/kg•°C)

Chapter Test B

Heat

MULTIPLE CHOICE

In the space provided, write the letter of the term or phrase that best completes each statement or best answers each question.

_____ 1. Which of the following is a direct cause of a substance's temperature increase?
  a. Energy is removed from the particles of the substance.
  b. Kinetic energy is added to the particles of the substance.
  c. The number of atoms and molecules in a substance changes.
  d. The volume of the substance decreases.

_____ 2. What is the temperature of a system in thermal equilibrium with another system made up of water and steam at 1 atm of pressure?
  a. 0°F
  b. 273 K
  c. 0 K
  d. 100°C

_____ 3. A substance registers a temperature change from 20°C to 40°C. To what incremental temperature change does this correspond?
  a. 20°F
  b. 40°F
  c. 36°F
  d. 313°F

_____ 4. What temperature has the same numerical value on both the Celsius and the Fahrenheit scales?
  a. −40°
  b. 0°
  c. 40°
  d. −72°

_____ 5. If energy is transferred from a table to a block of ice moving across the table, which of the following statements is true?
  a. The table and the ice are at thermal equilibrium.
  b. The ice is cooler than the table.
  c. The ice is no longer 0°C.
  d. Energy is being transferred from the ice to the table.

_____ 6. Energy is transferred as heat between two objects, one with a temperature of 5°C and the other with a temperature of 20°C. If two other objects are to have the same amount of energy transferred between them, what might their temperatures be?
  a. 10°C and 15°C
  b. 15°C and 25°C
  c. 17°C and 32°C
  d. 80°C and 90°C

❙ Chapter Test B *continued*

_____ **7.** In an elastic collision between two ball bearings, kinetic energy is conserved. If there is no change in potential energy, which of the following is true?

a. $\Delta U > 0$ **c.** $\Delta U < 0$

b. $\Delta U = 0$ **d.** ΔU cannot be determined for this event.

The figure below shows how the temperature of 10.0 g of ice changes as energy is added. Use the figure to answer questions 8–11.

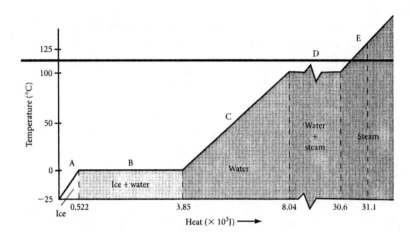

_____ **8.** Which of the following statements is correct?

a. The water absorbed energy continuously, but the temperature increased only when all of the water was in one phase.

b. The water absorbed energy sporadically, and the temperature increased only when all of the water was in one phase.

c. The water absorbed energy continuously, and the temperature increased continuously.

d. The water did not absorb energy.

_____ **9.** Which value equals the latent heat required to change the liquid water into steam?

a. 8.04×10^3 J   **c.** 30.6×10^3 J

b. 22.6×10^3 J   **d.** 31.1×10^3 J

_____ **10.** At what point on the figure above does the substance undergo a phase change?

a. A   **c.** C

b. B   **d.** E

_____11. If there is no temperature difference between a substance and its sur-
roundings, what has occurred on the microscopic level?
   a. Energy has been transferred from lower-energy particles to higher-
      energy particles.
   b. Energy has been transferred from higher-energy particles to lower-
      energy particles.
   c. No energy has been transferred between the substance and its
      surroundings.
   d. Heat has been flowing back and forth.

_____12. What three properties of a substance affect the amount of energy
transferred as heat to or from the substance?
   a. volume, temperature change, specific heat capacity
   b. density, temperature change, specific heat capacity
   c. mass, temperature change, specific heat capacity
   d. mass, temperature change, latent heat

_____13. A slice of bread contains about 4.19×10^5 J of energy. If the specific
heat capacity of a person is 4.19×10^3 J/kg•°C, by how many degrees
Celsius would the temperature of a 70.0 kg person increase if all the
energy in the bread were converted to heat?
   a. 2.25°C   c. 1.43°C
   b. 1.86°C   d. 1.00°C

SHORT ANSWER

14. Explain how thermal expansion makes a mercury thermometer a useful
device for measuring temperature.

_____

_____

_____

15. Is there energy transferred as heat between two objects that are in thermal
equilibrium? Explain your answer.

_____

_____

| Chapter Test B *continued*

16. Explain if energy can be transferred as heat from an object at a low temperature to an object at a high temperature, and if so, why objects don't become hotter spontaneously.

_____

_____

_____

_____

17. How does the principle of conservation of energy take internal energy into account?

_____

_____

_____

_____

PROBLEM

18. The surface temperature of Venus is 737 K. What is this temperature in degrees Fahrenheit?

19. A falling stone with a mass of 0.255 kg strikes the ground. Assuming that the stone is initially at rest when it begins falling, how high must the stone be above the ground for the internal energy of the stone and ground to increase by 2450 J? ($g = 9.81$ m/s^2)

20. A 0.10 kg piece of copper at an initial temperature of 95°C is dropped into 0.20 kg of water contained in a 0.28 kg aluminum calorimeter. The water and calorimeter are initially at 15°C. What is the final temperature of the system when it reaches equilibrium? ($c_{p,Cu} = 387$ J/kg•°C, $c_{p,Al} = 899$ J/kg•°C, and $c_{p,w} = 4186$ J/kg•°C)

Assessment

Chapter Test A

Thermodynamics
MULTIPLE CHOICE

In the space provided, write the letter of the term or phrase that best completes each statement or best answers each question.

_____ **1.** What accounts for an increase in the temperature of a gas that is kept at constant volume?
   a. Energy has been removed as heat from the gas.
   b. Energy has been added as heat to the gas.
   c. Energy has been removed as work done by the gas.
   d. Energy has been added as work done on the gas.

_____ **2.** An ideal gas system is maintained at a constant volume of 4 L. If the pressure is constant, how much work is done by the system?
   a. 0 J   **c.** 8 J
   b. 5 J   **d.** 30 J

_____ **3.** What thermodynamic process for an ideal gas system has an unchanging internal energy and a heat intake that corresponds to the value of the work done by the system?
   a. isovolumetric   **c.** adiabatic
   b. isobaric   **d.** isothermal

_____ **4.** Which thermodynamic process takes place when work is done on or by the system but no energy is transferred to or from the system as heat?
   a. isovolumetric   **c.** adiabatic
   b. isobaric   **d.** isothermal

_____ **5.** Which thermodynamic process takes place at a constant temperature so that the internal energy of a system remains unchanged?
   a. isovolumetric   **c.** adiabatic
   b. isobaric   **d.** isothermal

_____ **6.** According to the first law of thermodynamics, the difference between energy transferred to or from a system as heat and energy transferred to or from a system by work is equivalent to which of the following?
   a. entropy change   **c.** volume change
   b. internal energy change **d.** pressure change

_____ **7.** An ideal gas system undergoes an adiabatic process in which it expands and does 20 J of work on its environment. What is the change in the system's internal energy?
   a. −20 J   **c.** 0 J
   b. −5 J   **d.** 20 J

_____ **8.** An ideal gas system undergoes an adiabatic process in which it expands and does 20 J of work on its environment. How much energy is transferred to the system as heat?

    a. -20 J   **c.** 5 J

    b. 0 J   **d.** 20 J

_____ **9.** Which of the following is a thermodynamic process in which a system returns to the same conditions under which it started?

    a. an isovolumetric process   **c.** a cyclic process

    b. an isothermal process   **d.** an adiabatic process

_____ **10.** How does a real heat engine differ from an ideal cyclic heat engine?

    a. A real heat engine is not cyclic.

    b. An ideal heat engine converts all energy from heat to work.

    c. A real heat engine is not isolated, so matter enters and leaves the engine.

    d. An ideal heat engine is not isolated, so matter enters and leaves the engine.

_____ **11.** The requirement that a heat engine must give up some energy at a lower temperature in order to do work corresponds to which law of thermodynamics?

    a. first

    b. second

    c. third

    d. No law of thermodynamics applies.

_____ **12.** An ideal heat engine has an efficiency of 50 percent. Which of the following statements is *not* true?

    a. The amount of energy exhausted as heat equals the energy added to the engine as heat.

    b. The amount of work done is half the energy added to the engine as heat.

    c. The amount of energy exhausted as heat is half the energy added to the engine as heat.

    d. The amount of energy exhausted as heat equals the work done.

_____ **13.** What occurs when a system's disorder is increased?

    a. No work is done.

    b. No energy is available to do work.

    c. Less energy is available to do work.

    d. More energy is available to do work.

_____**14.** Imagine you could observe the individual atoms that make up a piece
of matter and that you observe the motion of the atoms becoming
more orderly. What can you assume about the system?
   a. It is gaining thermal energy.
   b. Its entropy is increasing.
   c. Its entropy is decreasing.
   d. Positive work is being done on the system.

SHORT ANSWER

15. A match is struck on a matchbook cover. How is energy transferred so that
the match can ignite and produce a flame?

_____

_____

_____

16. A mechanic pushes down very quickly on the plunger of an insulated pump.
The air hose is plugged so that no air escapes. What type of thermodynamic
process takes place? What type of energy transfer and change occurs?

_____

_____

_____

17. What is true of the internal energy of an isolated system?

_____

_____

18. According to the conservation of energy, what is true about the net work and
net heat in a cyclic process?

_____

_____

19. How does $Q_c > 0$ relate to the second law of thermodynamics?

_____

_____

_____

Chapter Test A *continued*

20. Explain why the efficiencies of real heat engines are always much less than the calculated maximum efficiencies of ideal heat engines.

_____

_____

_____

_____

21. What is entropy?

_____

_____

22. Why must work be done to reduce entropy in most systems?

_____

_____

_____

_____

PROBLEM

23. A container of gas is at a pressure of 3.7×10^5 Pa. How much work is done by the gas if its volume expands by 1.6 m^3?

24. A total of 165 J of work is done on a gaseous refrigerant as it undergoes compression. If the internal energy of the gas increases by 123 J during the process, what is the total amount of energy transferred as heat?

25. An engine adds 75 000 J of energy as heat and removes 15 000 J of energy as heat. What is the engine's efficiency?

Chapter Test B

Thermodynamics
MULTIPLE CHOICE

In the space provided, write the letter of the term or phrase that best completes each statement or best answers each question.

_____ **1.** When an ideal gas does positive work on its surroundings, which of the gas's quantities increases?
    a. temperature   **c.** pressure
    b. volume   **d.** internal energy

_____ **2.** Air cools as it escapes from a diver's compressed air tank. What kind of process is this?
    a. isovolumetric   **c.** adiabatic
    b. isobaric   **d.** isothermal

_____ **3.** In an isovolumetric process for an ideal gas, the system's change in the energy as heat is equivalent to a change in which of the following?
    a. temperature   **c.** pressure
    b. volume   **d.** internal energy

_____ **4.** How is conservation of internal energy expressed for a system during an adiabatic process?
    a. $Q = W = 0$, so $\Delta U = 0$ and $U_i = U_f$
    b. $Q = 0$, so $\Delta U = -W$
    c. $\Delta T = 0$, so $\Delta U = 0$; therefore, $\Delta U = Q - W = 0$, or $Q = W$
    d. $\Delta V = 0$, so $P\Delta V = 0$ and $W = 0$; therefore, $\Delta U = Q$

_____ **5.** An ideal gas system undergoes an isovolumetric process in which 20 J of energy is added as heat to the gas. What is the change in the system's internal energy?
    a. -20 J   **c.** 5 J
    b. 0 J   **d.** 20 J

_____ **6.** Which of the following is *not* a way in which a cyclic process resembles an isothermal process?
    a. Energy can be transferred as work.
    b. Energy can be transferred as heat.
    c. The temperature of the system remains constant throughout the process.
    d. There is no net change in the internal energy of the system.

_____ **7.** A heat engine has taken in energy as heat and used a portion of it to do work. What must happen next for the engine to complete the cycle and return to its initial conditions?

    a. It must give up energy as heat to a lower temperature so work can be done on it.

    b. It must give up energy as heat to a higher temperature so work can be done on it.

    c. It must do work to transfer the remaining energy as heat to a lower temperature.

    d. It must do work to transfer the remaining energy as heat to a higher temperature.

_____ **8.** An electrical power plant manages to transfer 88 percent of the heat produced in the burning of fossil fuel to convert water to steam. Of the heat carried by the steam, 40 percent is converted to the mechanical energy of the spinning turbine. Which best describes the overall efficiency of the heat-to-work conversion in the plant?

    a. greater than 88 percent   **c.** 40 percent

    b. 88 percent   **d.** less than 40 percent

_____ **9.** When a drop of ink mixes with water, what happens to the entropy of the system?

    a. The system's entropy increases, and the total entropy of the universe increases.

    b. The system's entropy decreases, and the total entropy of the universe increases.

    c. The system's entropy increases, and the total entropy of the universe decreases.

    d. The system's entropy decreases, and the total entropy of the universe decreases.

_____ **10.** A thermodynamic process occurs, and the entropy of a system decreases. What can be concluded about the entropy change of the environment?

    a. It decreases.

    b. It increases.

    c. It stays the same.

    d. It could increase or decrease, depending on the process.

SHORT ANSWER

11. A physics textbook is balanced on top of an inflated balloon on a cold morning. As the day passes, the temperature increases, the balloon expands, and the textbook rises. Is there a transfer of energy as heat? If so, what is it? Has any work been done? If so, on what?

_____

_____

12. A gas is confined in a cylinder with a piston. What happens when work is done on the gas?

_____

_____

13. What changes can be made to the transfer of energy as heat to a heat engine in order to increase the amount of work done by the engine?

_____

_____

_____

_____

14. Describe how energy is transferred as heat during the part of an engine cycle where the engine does work on the environment and during the part of the cycle when work is done on the engine.

_____

_____

_____

15. Use the second law of thermodynamics and the equation for heat engine efficiency to show why efficiency must always be less than 1.

_____

_____

_____

| Chapter Test B *continued*

16. Ice cubes are formed in the freezer compartment of a refrigerator. Explain the change in entropy of the water freezing, as well as the change in entropy of the environment outside the refrigerator. Does the water freezes spontaneously, and if not, why not?

_____

_____

_____

PROBLEM

17. An ideal gas is maintained at a constant pressure of 7.0×10^4 N/m^2 while its volume decreases by 0.20 m^3. What work is done by the system on its environment?

18. Over several cycles, a refrigerator compressor does work on the refrigerant. This work is equivalent to a constant pressure of 4.13×10^5 Pa compressing a circular piston with a radius of 0.019 m a distance of 25.0 m. If the change in the refrigerant's internal energy is 0 J after each cycle, how much heat will the refrigerant remove from within the refrigerator?

19. A steam engine takes in 2.06×10^5 J of energy added as heat and exhausts 1.53×10^5 J of energy removed as heat per cycle. What is its efficiency?

20. The gas within a cylinder of an engine undergoes a net change in volume of 1.50×10^{-3} m^3 when it does work at a constant pressure of 3.27×10^5 Pa. If the efficiency of the engine is 0.225, how much work must the engine give up as heat to the low-temperature reservoir?

Assessment

Chapter Test A

Vibrations and Waves
MULTIPLE CHOICE

In the space provided, write the letter of the term or phrase that best completes each statement or best answers each question.

_____ **1.** Which of the following is *not* an example of approximate simple harmonic motion?
  a. a ball bouncing on the floor
  b. a child swinging on a swing
  c. a piano wire that has been struck
  d. a car's radio antenna waving back and forth

_____ **2.** A mass attached to a spring vibrates back and forth. At the equilibrium position, the
  a. acceleration reaches a maximum.
  b. velocity reaches a maximum.
  c. net force reaches a maximum.
  d. velocity reaches zero.

_____ **3.** A simple pendulum swings in simple harmonic motion. At maximum displacement,
  a. the acceleration reaches a maximum.
  b. the velocity reaches a maximum.
  c. the acceleration reaches zero.
  d. the restoring force reaches zero.

_____ **4.** The angle between the string of a pendulum at its equilibrium position and at its maximum displacement is the pendulum's
  a. period.   **c.** vibration.
  b. frequency.   **d.** amplitude.

_____ **5.** For a mass hanging from a spring, the maximum displacement the spring is stretched or compressed from its equilibrium position is the system's
  a. amplitude.   **c.** frequency.
  b. period.   **d.** acceleration.

_____ **6.** For a system in simple harmonic motion, which of the following is the time required to complete a cycle of motion?
  a. amplitude   **c.** frequency
  b. period   **d.** revolution

_____ **7.** For a system in simple harmonic motion, which of the following is the
number of cycles or vibrations per unit of time?
a. amplitude   **c.** frequency
b. period   **d.** revolution

_____ **8.** Which of the following features of a given pendulum changes when the
pendulum is moved from Earth's surface to the moon?
a. the mass   **c.** the equilibrium position
b. the length   **d.** the restoring force

_____ **9.** A wave travels through a medium. As the wave passes, the particles of
the medium vibrate in a direction perpendicular to the direction of the
wave's motion. The wave is
a. longitudinal.   **c.** electromagnetic.
b. a pulse.   **d.** transverse.

_____ **10.** One end of a taut rope is fixed to a post. What type of wave is
produced if the free end is quickly raised and lowered one time?
a. pulse wave   **c.** sine wave
b. periodic wave   **d.** longitudinal wave

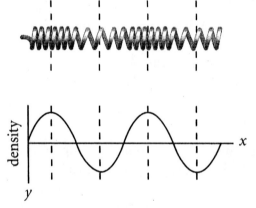

_____ **11.** Each compression in the waveform of the longitudinal wave shown
above corresponds to what feature of the transverse wave below it?
a. wavelength   **c.** troughs
b. crests   **d.** amplitude

_____ **12.** Which of the following most affects the wavelength of a mechanical
wave moving through a medium? Assume that the frequency of the
wave remains constant.
a. the nature of the medium   **c.** the height of a crest
b. the amplitude   **d.** the energy carried by the wave

_____**13.** When a mechanical wave's amplitude is tripled, the energy the wave
carries in a given time interval is increased by a factor of
a. 3.   **c.** 9.
b. 6.   **d.** 18.

_____**14.** Two mechanical waves can occupy the same space at the same time
because waves
a. are matter.
b. are displacements of matter.
c. do not cause interference patterns.
d. cannot pass through one another.

_____**15.** When two mechanical waves coincide, the amplitude of the resultant
wave is always _____ the amplitudes of each wave alone.
a. greater than   **c.** the sum of
b. less than   **d.** the same as

_____**16.** Waves arriving at a fixed boundary are
a. neither reflected nor inverted.   **c.** reflected and inverted.
b. reflected but not inverted.   **d.** inverted but not reflected.

_____**17.** Standing waves are produced by periodic waves of
a. any amplitude and wavelength traveling in the same direction.
b. the same amplitude and wavelength traveling in the same direction.
c. any amplitude and wavelength traveling in opposite directions.
d. the same frequency, amplitude, and wavelength traveling in
opposite directions.

_____**18.** How many nodes and antinodes are shown in the standing wave above?
a. two nodes and three antinodes
b. one node and two antinodes
c. one-third node and one antinode
d. three nodes and two antinodes

SHORT ANSWER

19. If a spring is stretched from a displacement of 10 cm to a displacement of
30 cm, the force exerted by the spring increases by a factor of _____.

_____

20. Two pulses of equal positive amplitude travel along a rope toward a fixed boundary. The first pulse is reflected and returns along the rope. When the two pulses meet and coincide, what kind of interference will occur? Explain.

_____

_____

_____

21. When two waves meet, they combine according to the _____ principle.

_____

22. Suppose that a pendulum has a period of 4.0 s at Earth's surface. If the pendulum is taken to the moon, where the acceleration due to gravity is much less than on Earth, will the pendulum's period increase, decrease, or stay the same? Explain your answer.

_____

_____

_____

_____

PROBLEM

23. If a force of 50 N stretches a spring 0.10 m, what is the spring constant?

24. An amusement park ride swings back and forth once every 40.0 s. What is the ride's frequency?

25. A periodic wave has a wavelength of 0.50 m and a speed of 20 m/s. What is the wave frequency?

Chapter Test B

Vibrations and Waves

MULTIPLE CHOICE

In the space provided, write the letter of the term or phrase that best completes each statement or best answers each question.

_____ **1.** Vibration of an object about an equilibrium point is called simple harmonic motion when the restoring force is proportional to
    a. time.
    b. displacement.
    c. a spring constant.
    d. mass.

_____ **2.** A mass-spring system can oscillate with simple harmonic motion because a compressed or stretched spring has what kind of energy?
    a. kinetic
    b. mechanical
    c. gravitational potential
    d. elastic potential

_____ **3.** A pendulum swings through a total of 28°. If the displacement is equal on each side of the equilibrium position, what is the amplitude of this vibration? (Disregard frictional forces acting on the pendulum.)
    a. 28°
    b. 14°
    c. 56°
    d. 7.0°

_____ **4.** If a pendulum is adjusted so that its frequency changes from 10 Hz to 20 Hz, its period will change from n seconds to
    a. $n/4$ seconds.
    b. $n/2$ seconds.
    c. $2n$ seconds.
    d. $4n$ seconds.

_____ **5.** By what factor should the length of a simple pendulum be changed in order to triple the period of vibration?
    a. 3
    b. 6
    c. 9
    d. 27

_____ **6.** Each stretched region in the waveform of the longitudinal wave shown on the right corresponds to what feature of the transverse wave below it?
    a. wavelength
    b. crests
    c. troughs
    d. amplitude

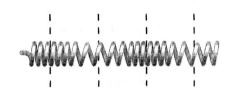

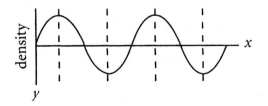

_____ **7.** Suppose that two sound waves passing through the same medium have different wavelengths. Which of the following is most likely to be the reason for the differing wavelengths?
  a. the nature of the medium   **c.** differences in frequency
  b. differences in amplitude   **d.** the type of wave

_____ **8.** When a mechanical wave's amplitude is reduced by half, the energy the wave carries in a given time interval is
  a. doubled.   **c.** decreased to one-half.
  b. increased by a factor of 1.4. **d.** decreased to one-fourth.

_____ **9.** Two waves traveling in opposite directions on a rope meet and undergo complete destructive interference. Which of the following best describes the waves a moment after the waves meet and coincide?
  a. The waves no longer exist.
  b. The waves continue unchanged.
  c. The waves reflect and travel backward.
  d. A single wave continues along the rope.

_____ **10.** Which of the following types of interference will occur when the pulses in the figure above meet?
  a. no interference   **c.** partial destructive
  b. complete constructive   **d.** complete destructive

_____ **11.** Consider two identical wave pulses on a rope having a fixed end. Suppose the first pulse reaches the end of the rope, is reflected back, and then meets the second pulse. When the two pulses overlap exactly, what will be the amplitude of the resultant pulse?
  a. zero
  b. same as the original pulses
  c. double the amplitude of the original pulses
  d. half the amplitude of the original pulses

_____ **12.** A 2.0 m long stretched rope is fixed at both ends. Which wavelength would *not* produce standing waves on this rope?
  a. 2.0 m   **c.** 4.0 m
  b. 3.0 m   **d.** 6.0 m

_____**13.** A 3.0 m long stretched string is fixed at both ends. If standing waves
with a wavelength of two-thirds L are produced on this string, how
many nodes will be formed?

    a. 0   **c.** 3

    b. 2   **d.** 4

SHORT ANSWER

14. In an oscillating mass-spring system, the restoring force is a result of the force
exerted by the spring. What causes the restoring force in a swinging pendulum?

_____

_____

_____

15. If a spring is stretched from a displacement of 10 cm to a displacement of 30 cm,
the force exerted by the spring increases by a factor of _____

_____

16. In an old-fashioned pendulum clock, the bob is moved up and down to adjust
the clock to keep accurate time. How would you adjust the bob in order to
correct a clock that runs too fast? Explain why the adjustment works.

_____

_____

_____

17. A boat produces a wave as it passes an aluminum can floating in a lake.
Explain why the can is not moved along in the direction of wave motion.

_____

_____

_____

18. What is the difference between a pulse wave and a periodic wave?

_____

_____

_____

19. What feature of a wave increases when the source of vibration increases in energy?

_____

20. A standing wave is produced by plucking a string. The points along the plucked string that appear not to be vibrating are produced by _____ interference.

_____

PROBLEM

21. A mass on a spring that has been compressed 0.1 m has a restoring force of 20 N. What is the spring constant?

22. A car with bad shock absorbers bounces up and down after hitting a bump. The car has a mass of 1500 kg and is supported by four springs, each having a spring constant of 6600 N/m. What is the period for each spring?

23. A student wishes to construct a mass-spring system that will oscillate with the same frequency as a swinging pendulum with a period of 3.45 s. The student has a spring with a spring constant of 72.0 N/m. What mass should the student use to construct the mass-spring system?

24. Radio waves from an FM station have a frequency of 103.1 MHz. If the waves travel with a speed of 3.00×10^8 m/s, what is the wavelength?

25. Vibration of a certain frequency produces a standing wave on a stretched string that is 2.0 m long. The standing wave has 7 nodes and 5 antinodes. What is the wavelength of the wave that produces this standing wave?

Assessment

Chapter Test A

Sound
MULTIPLE CHOICE

In the space provided, write the letter of the term or phrase that best completes each statement or best answers each question.

_____ **1.** Sound waves
   a. are a part of the electromagnetic spectrum.
   b. do not require a medium for transmission.
   c. are longitudinal waves.
   d. are transverse waves.

_____ **2.** The trough of the sine curve used to represent a sound wave corresponds to a
   a. compression.
   b. region of high pressure.
   c. point where molecules are pushed closer together.
   d. rarefaction.

_____ **3.** Which of the following is the region of a sound wave in which the density and pressure are greater than normal?
   a. rarefaction   **c.** amplitude
   b. compression   **d.** wavelength

_____ **4.** The highness or lowness of a sound is perceived as
   a. compression.   **c.** ultrasound.
   b. wavelength.   **d.** pitch.

_____ **5.** Pitch depends on the _____ of a sound wave.
   a. frequency   **c.** power
   b. amplitude   **d.** speed

_____ **6.** In general, sound travels faster through
   a. solids than through gases.
   b. gases than through solids.
   c. gases than through liquids.
   d. empty space than through matter.

_____ **7.** At a large distance from a sound source, spherical wave fronts are viewed as
   a. wavelengths.   **c.** rays.
   b. troughs.   **d.** plane waves.

_____ **8.** The distance between wave fronts of plane waves corresponds to
_____ of a sound wave.
- **a.** one wavelength
- **b.** two amplitudes
- **c.** one compression
- **d.** one rarefaction

_____ **9.** A train moves down the track toward an observer. The sound from the train, as heard by the observer, is _____ the sound heard by a passenger on the train.
- **a.** the same as
- **b.** a different timbre than
- **c.** higher in pitch than
- **d.** lower in pitch than

_____ **10.** The Doppler effect occurs with
- **a.** only sound waves.
- **b.** only transverse waves.
- **c.** only water waves.
- **d.** all waves.

_____ **11.** The property of sound called *intensity* is proportional to the rate at which energy flows through
- **a.** an area perpendicular to the direction of propagation.
- **b.** an area parallel to the direction of propagation.
- **c.** a cylindrical tube.
- **d.** a sound wave of a certain frequency.

_____ **12.** The perceived loudness of a sound is measured in
- **a.** hertz.
- **b.** decibels.
- **c.** watts.
- **d.** watts per square meter.

_____ **13.** Which of the following decibel levels is nearest to the value that you would expect for a running vacuum cleaner?
- **a.** 10 dB
- **b.** 30 dB
- **c.** 70 dB
- **d.** 120 dB

_____ **14.** A sound twice the intensity of the faintest audible sound is not perceived as twice as loud because the sensation of loudness in human hearing
- **a.** is approximately logarithmic.
- **b.** is approximately exponential.
- **c.** depends on the speed of sound.
- **d.** is proportional to frequency.

_____ **15.** When the frequency of a force applied to a system matches the natural frequency of vibration of the system, _____ occurs.
- **a.** damped vibration
- **b.** random vibration
- **c.** timbre
- **d.** resonance

_____16. When an air column vibrates in a pipe that is open at both ends,
  a. all harmonics are present.
  b. no harmonics are present.
  c. only odd harmonics are present.
  d. only even harmonics are present.

_____17. When an air column vibrates in a pipe that is closed at one end,
  a. all harmonics are present.
  b. no harmonics are present.
  c. only odd harmonics are present.
  d. only even harmonics are present.

_____18. The wavelength of the fundamental frequency of a vibrating string of
  length L is
  a. $1/2\ L$.   **c.** $2L$.
  b. L.   **d.** $4L$.

_____19. The quality of a musical tone of a certain pitch results from a
  combination of
  a. fundamental frequencies.   **c.** transverse waves.
  b. harmonics.   **d.** velocities.

_____20. Audible beats are formed by the interference of two waves
  a. of slightly different frequencies.
  b. of greatly different frequencies.
  c. with equal frequencies, but traveling in opposite directions.
  d. from the same vibrating source.

SHORT ANSWER

21. The region of a sound wave in which air molecules are pushed closer together
is called a(n) _____ .

_____

22. The _____ of a musical sound determines its pitch.

_____

23. What are the units used to express the intensity of a sound?

_____

| Chapter Test A *continued*

24. Under what conditions does sound resonance occur?

_____

_____

_____

PROBLEM

25. A wave on a guitar string has a velocity of 684 m/s. The guitar string is 62.5 cm long. What is the fundamental frequency of the vibrating string?

Assessment

Chapter Test B

Sound

MULTIPLE CHOICE

In the space provided, write the letter of the term or phrase that best completes each statement or best answers each question.

_____ **1.** Of the following materials, sound waves travel fastest through
  a. helium at 0°C.
  b. air at 0°C.
  c. copper at 0°C.
  d. air at 100°C.

_____ **2.** The point at which a ray crosses a wave front corresponds to a _____ of a sound wave.
  a. wavelength
  b. compression
  c. trough
  d. source

_____ **3.** A train moves down the track toward an observer. The sound from the train, as heard by the observer, is _____ the sound heard by a passenger on the train.
  a. the same as
  b. a different timbre than
  c. higher in pitch than
  d. lower in pitch than

_____ **4.** If you are on a train, how will the pitch of the train's whistle sound to you as the train moves?
  a. The pitch will become steadily higher.
  b. The pitch will become steadily lower.
  c. The pitch will not change.
  d. The pitch will become higher, and then become lower.

_____ **5.** At a distance of 3 m, the intensity of a sound will be _____ the intensity it was at a distance of 1 m.
  a. one-ninth
  b. one-third
  c. 3 times
  d. 9 times

_____ **6.** The intensity of a sound at any distance from the source is directly proportional to the sound's
  a. wavelength.
  b. pitch.
  c. power.
  d. frequency.

_____ **7.** If the intensity of a sound is increased by a factor of 100, the new decibel level will increase
  a. by two units.
  b. to twice the old one.
  c. by a factor of 10.
  d. by 20 units.

_____ **8.** The Tacoma Narrows bridge collapsed in 1940 when winds caused
_____ to build up in the bridge.
a. a compression wave   **c.** a standing wave
b. a longitudinal wave   **d.** friction

_____ **9.** For a standing wave in an air column in a pipe that is open at both
ends, there must be at least
a. one node and one antinode.
b. two nodes and one antinode.
c. two antinodes and one node.
d. two nodes and two antinodes.

_____ **10.** If a guitar string has a fundamental frequency of 500 Hz, what is the
frequency of its second harmonic?
a. 250 Hz   **c.** 1000 Hz
b. 750 Hz   **d.** 2000 Hz

_____ **11.** Musical instruments of different types playing the same note may often
be identified by the _____ of their sounds.
a. pitch   **c.** fundamental frequency
b. intensity   **d.** timbre

_____ **12.** How many beats per second are heard when two vibrating tuning forks
having frequencies of 342 Hz and 345 Hz are held side by side?
a. 687 Hz   **c.** 5 Hz
b. 343.5 Hz   **d.** 3 Hz

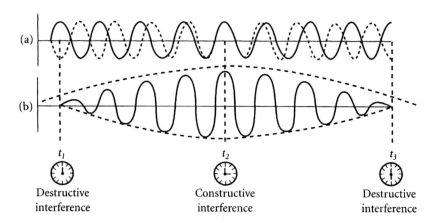

t_1	t_2	t_3
Destructive interference	Constructive interference	Destructive interference

_____ **13.** In the figure shown above, a beat occurs at
a. t_1.   **c.** t_3.
b. t_2.   **d.** t_1 and t_3.

SHORT ANSWER

14. Unlike a transverse wave on a rope, sound travels as a(n) _____
wave.

_____

15. Each trough of a sine wave used to represent a sound wave corresponds to
a(n) _____ of the sound wave.

_____

16. What happens to pitch when the frequency of a sound wave increases?

_____

17. Under what conditions can a spherical wave front be regarded as a plane
wave front? Explain.

_____

_____

_____

_____

18. Describe any changes in pitch of the sound a stationary observer hears from
the siren of an ambulance as the ambulance passes the observer.

_____

_____

_____

19. What happens to the intensity of a sound as the distance from the source
increases from 10 m to 20 m? Explain.

_____

_____

_____

20. Suppose you play a music CD in a room where there is a piano. During the
silence between songs, you hear a musical note coming from the piano, but
no one has touched the piano. Explain your observation.

_____

_____

| Chapter Test B *continued*

21. It is possible for a highly amplified musical note to cause a crystal goblet to shatter. Explain how this might occur.

_____

_____

_____

PROBLEM

22. What is the intensity of sound waves produced by a trumpet at a distance of 1.6 m when the power output of the trumpet is 0.30 W?

23. What length of guitar string would vibrate at a fundamental frequency of 825 Hz if the string is stretched so that the velocity of waves on the string is 577 m/s?

24. Closed-end organ pipes often have a movable closure that can be adjusted up or down to tune the pipe to the desired pitch. The resonant length of a certain closed-end pipe is 27.5 cm. By how many centimeters must this pipe be shortened or lengthened in order to tune it to a fundamental frequency of 349 Hz? The pipe is in an environment in which the speed of sound is 348 m/s.

25. As two notes are sounded, 6 beats per second are heard. The frequency of one note is 571 Hz. What is the frequency of the other note?

Assessment

Chapter Test A

Light and Reflection
MULTIPLE CHOICE

In the space provided, write the letter of the term or phrase that best completes each statement or best answers each question.

_____ **1.** Which portion of the electromagnetic spectrum is used in a television?
a. infrared waves   **c.** radio waves
b. X rays   **d.** gamma waves

_____ **2.** Which portion of the electromagnetic spectrum is used in a microscope?
a. infrared waves   **c.** visible light
b. gamma rays   **d.** ultraviolet light

_____ **3.** Which portion of the electromagnetic spectrum is used to identify fluorescent minerals?
a. ultraviolet light   **c.** infrared waves
b. X rays   **d.** gamma rays

_____ **4.** In a vacuum, electromagnetic radiation of short wavelengths
a. travels as fast as radiation of long wavelengths.
b. travels slower than radiation of long wavelengths.
c. travels faster than radiation of long wavelengths.
d. can travel both faster and slower than radiation of long wavelengths.

_____ **5.** If you know the wavelength of any form of electromagnetic radiation, you can determine its frequency because
a. all wavelengths travel at the same speed.
b. the speed of light varies for each form.
c. wavelength and frequency are equal.
d. the speed of light increases as wavelength increases.

_____ **6.** The farther light is from a source,
a. the more spread out light becomes.
b. the more condensed light becomes.
c. the more bright light becomes.
d. the more light is available per unit area.

_____ **7.** A highly polished finish on a new car provides a _____ surface for _____ reflection.
a. rough, diffused   **c.** rough, regular
b. specular, diffused   **d.** smooth, specular

_____ **8.** When a straight line is drawn perpendicular to a flat mirror at the point where an incoming ray strikes the mirror's surface, the angles of incidence and reflection are measured from the normal and
   a. the angles of incidence and reflection are equal.
   b. the angle of incidence is greater than the angle of reflection.
   c. the angle of incidence is less than the angle of reflection.
   d. the angle of incidence can be greater than or less than the angle of reflection.

_____ **9.** The image of an object in a flat mirror is always
   a. larger than the object.
   b. smaller than the object.
   c. independent of the size of the object.
   d. the same size as the object.

_____ **10.** Which of the following best describes the image produced by a flat mirror?
   a. virtual, inverted, and magnification greater than one
   b. real, inverted, and magnification less than one
   c. virtual, upright, and magnification equal to one
   d. real, upright, and magnification equal to one

_____ **11.** When the reflection of an object is seen in a flat mirror, the distance from the mirror to the image depends on
   a. the wavelength of light used for viewing.
   b. the distance from the object to the mirror.
   c. the distance of both the observer and the object to the mirror.
   d. the size of the object.

_____ **12.** What type of mirror is used whenever a magnified image of an object is needed?
   a. flat mirror   **c.** convex mirror
   b. concave mirror   **d.** two-way mirror

_____ **13.** The mirror equation and ray diagrams are valid concepts only for what type of rays?
   a. parallel rays   **c.** intersecting rays
   b. perpendicular rays   **d.** paraxial rays

_____ **14.** Object distance, image distance, and radius of curvature are _____ for curved mirrors.
   a. interdependent   **c.** directly related
   b. independent   **d.** unrelated

_____15. For a spherical mirror, the focal length is equal to _____ the radius of curvature of the mirror.
   a. one-fourth
   b. one-third
   c. one-half
   d. the square of

_____16. A parabolic mirror, instead of a spherical mirror, can be used to reduce the occurrence of which effect?
   a. spherical aberration
   b. mirages
   c. chromatic aberration
   d. light scattering

_____17. When red light and green light shine on the same place on a piece of white paper, the spot appears to be
   a. yellow.
   b. brown.
   c. white.
   d. black.

_____18. Which of the following is *not* an additive primary color?
   a. yellow
   b. blue
   c. red
   d. green

_____19. Which of the following is *not* a primary subtractive color?
   a. yellow
   b. cyan
   c. magenta
   d. blue

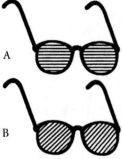
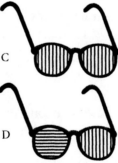

_____20. Which pair of glasses shown above is best suited for automobile drivers? The transmission axes are shown by straight lines on the lenses. (Hint: The light reflects off the hood of the car.)
   a. A
   b. B
   c. C
   d. D

|Chapter Test A *continued*

SHORT ANSWER

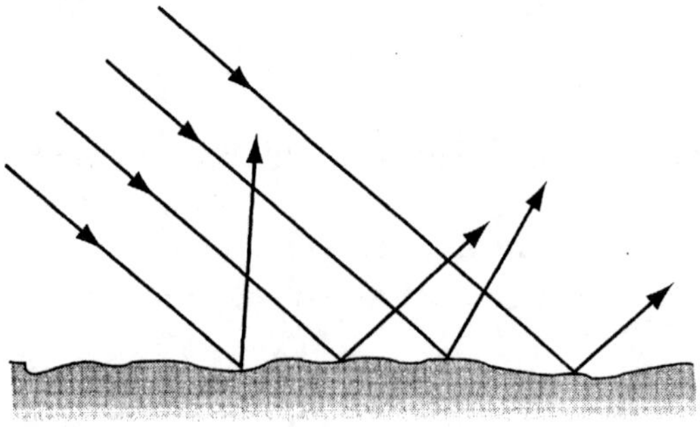

21. What type of reflection is illustrated in the figure shown above?

_____

_____

22. A line parallel to the principal axis is drawn from the object to a spherical mirror. How should the reflected ray be drawn?

_____

_____

23. What type of image do flat mirrors always form?

_____

_____

24. What percentage of light passes through a polarizing filter when the transmission axis is perpendicular to the plane of polarization for light?

_____

_____

PROBLEM

25. Yellow-green light has a wavelength of 560 nm. What is its frequency?

Name _____ Class _____ Date _____

Chapter Test B

Light and Reflection
MULTIPLE CHOICE

In the space provided, write the letter of the term or phrase that best completes each statement or best answers each question.

_____ 1. What is the wavelength of microwaves of 3.0×10^9 Hz frequency?
   a. 0.050 m   **c.** 0.10 m
   b. 0.060 m   **d.** 0.20 m

_____ 2. What is the frequency of an electromagnetic wave with a wavelength of 1.0×10^5 m?
   a. 1.0×10^{13} Hz   **c.** 3.0×10^{13} Hz
   b. 3.0×10^3 Hz   **d.** 1.0×10^3 Hz

_____ 3. When red light is compared with violet light,
   a. both have the same frequency.
   b. both have the same wavelength.
   c. both travel at the same speed.
   d. red light travels faster than violet light.

_____ 4. The relationship between frequency, wavelength, and speed holds for light waves because
   a. light travels slower in a vacuum than in air.
   b. all forms of electromagnetic radiation travel at a single speed in a vacuum.
   c. light travels in straight lines.
   d. different forms of electromagnetic radiation travel at different speeds.

_____ 5. If you are reading a book and you move twice as far away from the light source, how does the brightness at the new distance compare with that at the old distance? It is
   a. one-eighth as bright.   **c.** one-half as bright.
   b. one-fourth as bright.   **d.** twice as bright.

_____ 6. Snow reflects almost all of the light incident upon it. However, a single beam of light is not reflected in the form of parallel rays. This is an example of _____ reflection off a _____ surface.
   a. regular, rough   **c.** diffuse, specular
   b. regular, specular   **d.** diffuse, rough

_____ **7.** When incoming rays of light strike a flat mirror at an angle close to the surface of the mirror, the reflected rays are
   a. inclined high above the mirror's surface.
   b. parallel to the mirror's surface.
   c. perpendicular to the mirror's surface.
   d. close to the mirror's surface.

_____ **8.** If a light ray strikes a flat mirror at an angle of 29° from the normal, the reflected ray will be
   a. 29° from the normal.   **c.** 29° from the mirror's surface.
   b. 27° from the normal.   **d.** 61° from the normal.

_____ **9.** When two parallel mirrors are placed so that their reflective sides face each other, _____ images form. This is because the image in one mirror becomes the _____ for the other mirror.
   a. multiple, object   **c.** inverted, center of curvature
   b. reduced, virtual image   **d.** enlarged, focal point

_____ **10.** A concave mirror with a focal length of 10.0 cm creates a real image 30.0 cm away on its principal axis. How far from the mirror is the corresponding object?
   a. 20 cm   **c.** 7.5 cm
   b. 15 cm   **d.** 5.0 cm

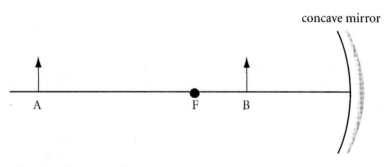

_____ **11.** In the diagram shown above, the image of object B would be
   a. virtual, enlarged, and inverted.
   b. real, enlarged, and upright.
   c. virtual, reduced, and upright.
   d. virtual, enlarged, and upright.

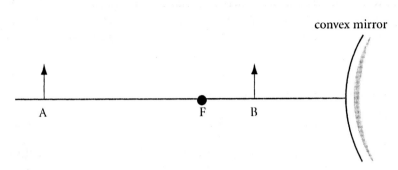

convex mirror

_____12. In the diagram shown above, the image of object B would be
     a. real, reduced, and upright.
     b. virtual, enlarged, and upright.
     c. virtual, reduced, and inverted.
     d. virtual, reduced, and upright.

_____13. When parallel rays that are also parallel to the principal axis strike
     a spherical mirror, rays that strike the mirror _____ the
     principal axis are focused at the focal point. Those rays that strike
     the mirror _____ the principal axis are focused at points
     between the mirror and the focal point.
     a. perpendicular to, far from
     b. close to, perpendicular to
     c. close to, far from
     d. far from, close to

_____14. If you looked at a light through the lenses from two polarizing sun-
     glasses that were overlapped at right angles to each other,
     a. all of the light would pass through.
     b. most of the light would pass through.
     c. little of the light would pass through.
     d. none of the light would pass through.

SHORT ANSWER

15. Why is the ultraviolet portion of the electromagnetic spectrum, not the
microwave or infrared red portion, used to sterilize medical instruments?

_____

_____

_____

| **Chapter Test B** *continued*

16. The electromagnetic spectrum is continuous and there is no sharp division between one kind of wave and the next. Assuming that the preceding statement is true, how do physicists distinguish one type of electromagnetic wave from another?

_____

_____

17. How are the terms *luminous flux* and *illuminance* related to each other?

_____

_____

_____

_____

18. If a light ray strikes a flat mirror with an angle of incidence equal to 52°, what is the angle of reflection? Explain.

_____

_____

PROBLEM

19. A candle 15 cm high is placed in front of a concave mirror at the focal point. The radius of curvature is 60 cm. Draw a ray diagram to determine the position and magnification of the image.

20. An object that is 2.00 cm high is placed 10.0 cm in front of a concave mirror with a radius of curvature of 40.0 cm. Find the magnification and location of the corresponding image in relation to the mirror's surface. Draw a ray diagram to confirm the position and magnification of the image.

Chapter Test A

Refraction

MULTIPLE CHOICE

In the space provided, write the letter of the term or phrase that best completes each statement or best answers each question.

_____ **1.** Part of a pencil that is placed in a glass of water appears bent in relation to the part of the pencil that extends out of the water. What is this phenomenon called?
  a. interference
  b. refraction
  c. diffraction
  d. reflection

_____ **2.** Refraction is the bending of a wave disturbance as it passes at an angle from one _____ into another.
  a. glass
  b. medium
  c. area
  d. boundary

_____ **3.** The _____ of light can change when light is refracted because the medium changes.
  a. frequency
  b. color
  c. speed
  d. transparency

_____ **4.** Light is *not* refracted when it is
  a. traveling from air into a glass of water at an angle of 35° to the normal.
  b. traveling from water into air at an angle of 35° to the normal.
  c. striking a wood surface at an angle of 75°.
  d. traveling from air into a diamond at an angle of 45°.

_____ **5.** When light passes at an angle to the normal from one material into another material in which its speed is higher,
  a. it is bent toward the normal to the surface.
  b. it always lies along the normal to the surface.
  c. it is unaffected.
  d. it is bent away from the normal to the surface.

_____ **6.** When light passes at an angle to the normal from one material into another material in which its speed is lower,
  a. it is bent toward the normal to the surface.
  b. it always lies along the normal to the surface.
  c. it is unaffected.
  d. it is bent away from the normal to the surface.

_____ **7.** What type of image is formed when rays of light actually intersect?
  a. real
  b. virtual
  c. curved
  d. projected

_____ **8.** What type of image does a converging lens produce?
    a. real   **c.** real or virtual
    b. virtual   **d.** none of the above

_____ **9.** In what direction does a parallel ray from an object proceed after passing through a diverging lens?
    a. The ray passes through the center of curvature, C.
    b. The ray continues parallel to the principal axis.
    c. The ray passes through the center of the lens.
    d. The ray is directed away from the focal point, F.

_____ **10.** In what direction does a focal ray from an object proceed after passing through a converging lens?
    a. The ray passes through the focal point, F.
    b. The ray passes through the center of the lens.
    c. The ray exits the lens parallel to the principal axis.
    d. The ray intersects with the center of curvature, C.

_____ **11.** In what direction does a focal ray from an object proceed after passing through a diverging lens?
    a. The ray passes through the focal point, F.
    b. The ray passes through the center of the lens.
    c. The ray exits the lens parallel to the principal axis.
    d. The ray intersects with the center of curvature, C.

_____ **12.** In what direction does a parallel ray from an object proceed after passing through a converging lens?
    a. The ray passes through the focal point, F.
    b. The ray continues parallel to the principal axis.
    c. The ray passes through the center of the lens.
    d. The ray is directed away from the focal point, F.

_____ **13.** How many focal points and focal lengths do converging and diverging lenses have?
    a. two, one   **c.** one, one
    b. one, two   **d.** two, two

_____ **14.** The focal length for a converging lens is
    a. always positive.
    b. always negative.
    c. dependent on the location of the object.
    d. dependent on the location of the image.

_____ **15.** A virtual image has a _____ image distance (q) and is located in _____ of the lens.
    a. positive, front   **c.** negative, front
    b. positive, back   **d.** negative, back

_____**16.** The focal length for a diverging lens is
    a. always positive.
    b. always negative.
    c. dependent on the location of the object.
    d. dependent on the location of the image.

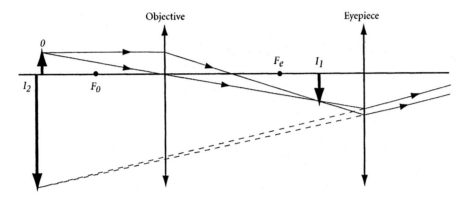

_____**17.** In the diagram of a compound microscope shown above, where would
    you place the slide?
    a. at O **c.** at F_0
    b. at I_2 **d.** at I_1

_____**18.** Which of the following describes what will happen to a light ray
    incident on a glass-to-air boundary at greater than the critical angle?
    a. total internal reflection
    b. total external transmission
    c. partial reflection, partial transmission
    d. partial reflection, total transmission

_____**19.** Atmospheric refraction of light rays is responsible for which of the
    following effects?
    a. spherical aberration
    b. mirages
    c. chromatic aberration
    d. total internal reflection in a gemstone

_____**20.** Which is *not* correct when describing the formation of rainbows?
    a. A rainbow is really spherical in nature.
    b. Sunlight is spread into a spectrum when it enters a spherical
    raindrop.
    c. Sunlight is internally reflected on the back side of a raindrop.
    d. All wavelengths refract at the same angle.

SHORT ANSWER

21. What happens to the speed of light as it moves into a substance with a higher index of refraction?

_____

_____

22. What does a positive magnification signify?

_____

_____

23. What condition must be met before total internal refraction can occur?

_____

_____

_____

24. How does white light passing through a prism produce a visible spectrum?

_____

_____

_____

PROBLEM

25. A ray of light passes from air into carbon disulfide ($n = 1.63$) at an angle of 28.0° to the normal. What is the refracted angle?

Chapter Test B

Refraction

MULTIPLE CHOICE

In the space provided, write the letter of the term or phrase that best completes each statement or best answers each question.

_____ 1. Which is an example of refraction?
   a. A parabolic mirror in a headlight focuses light into a beam.
   b. A fish appears closer to the surface of the water than it really is when observed from a riverbank.
   c. In a mirror, when you lift your right arm, the left arm of your image is raised.
   d. Light is bent slightly around corners.

_____ 2. The _____ of light can change when light is refracted because the velocity changes.
   a. frequency   **c.** wavelength
   b. medium   **d.** transparency

_____ 3. When a light ray moves from air into glass, which has a higher index of refraction, its path is
   a. bent toward the normal.
   b. bent away from the normal.
   c. parallel to the normal.
   d. not bent.

_____ 4. When a light ray passes from water ($n = 1.333$) into diamond ($n = 2.419$) at an angle of 45°, its path is
   a. bent toward the normal.
   b. bent away from the normal.
   c. parallel to the normal.
   d. not bent.

_____ 5. When a light ray passes from zircon ($n = 1.923$) into fluorite ($n = 1.434$) at an angle of 60°, its path is
   a. bent toward the normal.   **c.** parallel to the normal.
   b. bent away from the normal.   **d.** not bent.

_____ 6. A ray of light in air is incident on an air-to-glass boundary at an angle of exactly $3.0 \times 10^{1\circ}$ with the normal. If the index of refraction of the glass is 1.65, what is the angle of the refracted ray within the glass with respect to the normal?
   a. 56°   **c.** 29°
   b. 46°   **d.** 18°

_____ **7.** All of the following images can be formed by a converging lens *except* which one?

    a. virtual, upright, and magnified

    b. real and point

    c. real, inverted, and magnified

    d. real, upright, and magnified

_____ **8.** An object is placed 20.0 cm from a thin converging lens along the axis of the lens. If a real image forms behind the lens at a distance of 8.00 cm from the lens, what is the focal length of the lens?

    a. 5.71 cm   **c.** -13.3 cm

    b. 12.0 cm   **d.** 13.3 cm

_____ **9.** A film projector produces a 1.51 m image of a horse on a screen. If the projector lens is 4.00 m from the screen and the size of the horse on the film is 1.07 cm, what is the magnification of the image?

    a. 141   **c.** 7.08×10^{-3}

    b. -14.1 **d.** -7.08×10^{-3}

_____ **10.** If atmospheric refraction did not occur, how would the apparent time of sunrise and sunset be changed?

    a. Both would be later.

    b. Both would be earlier.

    c. Sunrise would be later, and sunset would be earlier.

    d. Sunrise would be earlier, and sunset would be later.

SHORT ANSWER

11. What type of situation will produce the largest amount of bending when a light ray crosses the boundary between two transparent media?

_____

_____

12. How are two converging lenses used to view an object in a compound microscope?

_____

_____

13. Why is it impossible to see an atom with a compound microscope?

_____

_____

| Chapter Test B *continued*

14. No refraction occurs when a light ray that is parallel to the normal strikes a transparent medium. Use the wave model of light to explain why.

_____

_____

_____

_____

_____

_____

15. What is the position and kind of image produced by the lens shown below? Draw a ray diagram to support your answer.

_____

_____

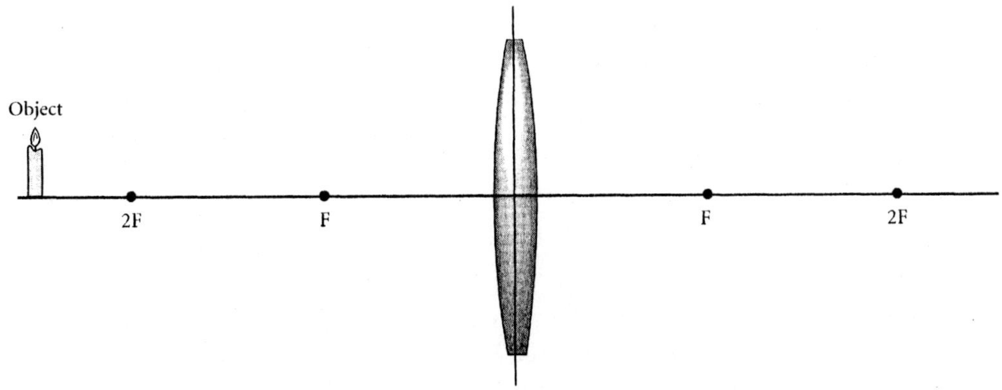

16. The critical angle for internal reflection inside a certain transparent material is found to be 48°. If entering light strikes the transparent material with an angle of incidence of 52°, predict how the light will be refracted.

_____

_____

17. Why are we able to see the sun in the morning before it actually rises above the horizon?

_____

_____

_____

| **Chapter Test B** *continued*

18. What is dispersion?

_____

_____

_____

_____

PROBLEM

19. An object is placed along the principal axis of a thin converging lens that has a focal length of 16 cm. If the distance from the object to the lens is 24 cm, what is the distance from the image to the lens?

20. The objective lens of a compound microscope has a focal length of 1.00 cm. A specimen is 1.25 cm from the objective lens. The image formed by the objective lens is 0.180 cm inside the focal point of the eyepiece whose focal length is 1.50 cm. What is the distance from the eyepiece to the image formed by the eyepiece lens?

Chapter Test A

Interference and Diffraction
MULTIPLE CHOICE

In the space provided, write the letter of the term or phrase that best completes each statement or best answers each question.

_____ **1.** In a double-slit interference pattern, the path length from one slit to the first dark fringe is longer than the path length from the other slit to the fringe by
   a. three-quarters of a wavelength.
   b. one-half of a wavelength.
   c. one-quarter of a wavelength.
   d. one full wavelength.

_____ **2.** In a double-slit interference experiment, a wave from one slit arrives at a point on a screen one wavelength behind the wave from the other slit. What is observed at that point?
   a. dark fringe
   b. bright fringe
   c. multicolored fringe
   d. gray fringe, neither dark nor bright

_____ **3.** In a double-slit interference experiment, a wave from one slit arrives at a point on a screen one-half wavelength behind the wave from the other slit. What is observed at that point?
   a. dark fringe
   b. bright fringe
   c. multicolored fringe
   d. gray fringe, neither dark nor bright

_____ **4.** If two lightbulbs are placed side by side, no interference is observed because
   a. each bulb produces many wavelengths of light.
   b. each bulb produces only one wavelength of light.
   c. incandescent light is incoherent.
   d. incandescent light is coherent.

_____ **5.** Coherence is the property by which two waves with identical wavelengths maintain a constant
   a. amplitude.   **c.** phase relationship.
   b. frequency.   **d.** speed.

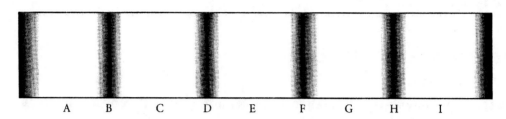

A   B   C   D   E   F   G   H   I

_____ **6.** The figure above shows the pattern of a double-slit interference experiment. The center of the pattern is located at E. Which fringe represents a second-order minimum?

    a. E   **c.** G

    b. F   **d.** H

_____ **7.** To produce a sustained interference pattern by light waves from multiple sources, which condition or conditions must be met?

    a. Sources must be coherent.

    b. Sources must be monochromatic.

    c. Sources must be coherent and monochromatic.

    d. Sources must be neither coherent nor monochromatic.

_____ **8.** Two beams of coherent light are shining on the same sheet of white paper. When referring to the crests and troughs of such waves, where will darkness appear on the paper?

    a. where the crest from one wave overlaps the crest from the other

    b. where the crest from one wave overlaps the trough from the other

    c. where the troughs from both waves overlap

    d. Darkness cannot occur because the two waves are coherent.

_____ **9.** For high resolution in optical instruments, the angle between resolved objects should be

    a. as small as possible.

    b. as large as possible.

    c. $1.22°$.

    d. $45°$.

_____ **10.** If light waves are coherent,

    a. they shift over time.

    b. their intensity is less than that of incoherent light.

    c. they remain in phase.

    d. they have less than three different wavelengths.

_____ **11.** In a laser, energy is added to a(n)

    a. mirror.

    b. active medium.

    c. partially transparent mirror.

    d. light wave.

_____**12.** Which of the following is the process of using a light wave to produce
more waves with properties identical to those of the first wave?
a. stimulated emission
b. active medium
c. hologram
d. bandwidth

_____**13.** Which of the following is a device that produces an intense, nearly
parallel beam of coherent light?
a. spectroscope
b. telescope
c. laser
d. diffraction grating

_____**14.** The acronym *laser* stands for light amplification by _____
emission of radiation.
a. similar
b. simultaneous
c. spontaneous
d. stimulated

_____**15.** A laser can be used
a. to treat glaucoma.
b. to measure distance.
c. to read bar codes.
d. All of the above

SHORT ANSWER

16. What is diffraction?

_____

_____

_____

17. What instrument uses a diffraction grating to separate light from a source into
its monochromatic components?

_____

18. What is the function of a spectrometer?

_____

_____

19. What is resolving power?

_____

_____

PROBLEM

20. The distance between the two slits in a double-slit experiment is 0.0025 mm. The third-order bright fringe ($m = 3$) is measured on a screen at an angle of $35°$ from the central maximum. What is the wavelength of the light?

Chapter Test B

Interference and Diffraction

MULTIPLE CHOICE

In the space provided, write the letter of the term or phrase that best completes each statement or best answers each question.

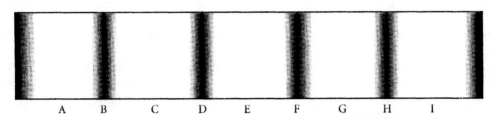

The figure above shows the pattern of a double-slit interference experiment. The center of the pattern is located at E. Use the figure to answer questions 1–4.

_____ 1. For which of the following fringes is the path length of the light wave from one slit more than one wavelength greater than the path length of the light wave from the other slit?
   a. A and I
   b. A, B, and C
   c. B and C
   d. B, C, D, and E

_____ 2. θ_1 is the angle between the central maximum and the first-order maximum. What is the angle between fringes D and F?
   a. $\dfrac{\theta_1}{2}$
   b. θ_1
   c. $\dfrac{3\theta_1}{2}$
   d. $2\theta_1$

_____ 3. In the experiment, the same slits are illuminated with light of greater wavelength. Which of the following would occur to the pattern shown?
   a. E would shift to the right.
   b. E would shift to the left.
   c. F would shift to the right.
   d. F would shift to the left.

_____ 4. In the experiment, the two slits are moved closer. Which of the following would occur to the pattern shown?
   a. E would shift to the right.
   b. E would shift to the left.
   c. F would shift to the right.
   d. F would shift to the left.

_____ **5.** Interference effects observed in the early 19th century were instrumental in supporting a concept of the existence of which property of light?

    a. polarization   **c.** wave nature

    b. particle nature   **d.** electromagnetic character

_____ **6.** For stable interference to occur, the phase difference must be

    a. incoherent.   **c.** $\frac{1}{2}\lambda$.

    b. monochromatic.   **d.** constant.

_____ **7.** The distance between the two slits in a double-slit interference experiment is 0.040 mm. The second-order bright fringe ($m = 2$) is measured on a screen at an angle of 2.2° from the central maximum. What is the wavelength of the light?

    a. 560 nm   **c.** 750 nm

    b. 630 nm   **d.** 770 nm

_____ **8.** The distance between two slits in a double-slit interference experiment is 0.0050 mm. What is the angle of the third-order bright fringe ($m = 3$) produced with light of 550 nm?

    a. 5.0°   **c.** 12°

    b. 9.9°   **d.** 19°

_____ **9.** At the first dark band in a single-slit diffraction pattern, the path lengths of selected pairs of wavelets differ by

    a. one wavelength.

    b. more than one wavelength.

    c. one-half wavelength.

    d. less than half of one wavelength.

_____ **10.** Monochromatic light shines on the surface of a diffraction grating with 5.3×10^3 lines/cm. The first-order maximum is observed at an angle of 17°. Find the wavelength.

    a. 420 nm   **c.** 530 nm

    b. 520 nm   **d.** 550 nm

_____ **11.** Light with a wavelength of 400.0 nm passes through a 1.00×10^4 lines/cm diffraction grating. What is the second-order angle of diffraction?

    a. 21.3°   **c.** 56.5°

    b. 53.1°   **d.** 72.1°

_____ **12.** The angle between the first-order maximum and the central maximum for monochromatic light of 2300 nm is 27°. Calculate the number of lines per centimeter on this grating.

    a. 1600 lines/cm   **c.** 2500 lines/cm

    b. 2.0×10^3 lines/cm   **d.** 4500 lines/cm

| Chapter Test B *continued*

SHORT ANSWER

13. Describe the pattern that results from the single-slit diffraction of monochromatic light.

_____

_____

_____

14. Why is the resolving power for optical telescopes on Earth limited?

_____

_____

15. How does an increase in the wavelength of light reaching an optical instrument with a set aperture affect the resolving power of the instrument?

_____

_____

16. What is meant by the statement that a laser produces a narrow beam of coherent light?

_____

_____

_____

17. How does a laser produce coherent light?

_____

_____

_____

_____

_____

_____

_____

PROBLEM

18. The distance between two slits in a double-slit experiment is 4.2×10^{-6} m. The first-order bright fringe is measured on a screen at an angle of 8.0° from the central maximum. What is the wavelength of the light?

19. Monochromatic light shines on the surface of a diffraction grating with 6.0×10^3 lines/cm. The angle between the central maximum and the first dark fringe is 10.0°. Find the wavelength.

20. Monochromatic light from a helium-neon laser ($\lambda = 632.8$ nm) shines at a right angle onto the surface of a diffraction grating that contains 630 692 lines/m. At what angles would you observe the first-order and second-order maxima?

Assessment

Chapter Test A

Electric Forces and Fields
MULTIPLE CHOICE
In the space provided, write the letter of the term or phrase that best completes each statement or best answers each question.

_____ **1.** What happens when a rubber rod is rubbed with a piece of fur, giving it a negative charge?
    a. Protons are removed from the rod.
    b. Electrons are added to the rod.
    c. Electrons are added to the fur.
    d. The fur is left neutral.

_____ **2.** A repelling force occurs between two charged objects when the charges are of
    a. unlike signs.   **c.** equal magnitude.
    b. like signs.   **d.** unequal magnitude.

_____ **3.** An attracting force occurs between two charged objects when the charges are of
    a. unlike signs.   **c.** equal magnitude.
    b. like signs.   **d.** unequal magnitude.

_____ **4.** When a glass rod is rubbed with silk and becomes positively charged,
    a. electrons are removed from the rod.
    b. protons are removed from the silk.
    c. protons are added to the silk.
    d. the silk remains neutral.

_____ **5.** Electric charge is
    a. found only in a conductor.   **c.** found only in insulators.
    b. conserved.   **d.** not conserved.

_____ **6.** Charge is most easily transferred in
    a. nonconductors.   **c.** semiconductors.
    b. conductors.   **d.** insulators.

_____ **7.** The process of charging a conductor by bringing it near another charged object and then grounding the conductor is called
    a. contact charging.   **c.** polarization
    b. induction.   **d.** neutralization.

_____ **8.** The figure shown on the right demonstrates charging by
  a. grounding. **c.** polarization.
  b. induction. **d.** contact.

_____ **9.** Both insulators and conductors can be charged by
  a. grounding. **c.** polarization.
  b. induction. **d.** contact.

_____**10.** A surface charge can be produced on insulators by
  a. grounding. **c.** polarization.
  b. induction. **d.** contact.

Charged object

Insulator

Induced charges

_____**11.** Conductors can be charged by _____, while insulators cannot.
  a. grounding **c.** polarization
  b. induction **d.** contact

_____**12.** Which of the following is *not* true for both gravitational and electric forces?
  a. The inverse square distance law applies.
  b. Forces are proportional to physical properties.
  c. Potential energy is a function of distance of separation.
  d. Forces are either attractive or repulsive.

_____**13.** Electric field strength depends on
  a. charge and distance.
  b. charge and mass.
  c. Coulomb constant and mass.
  d. elementary charge and radius.

_____**14.** What occurs when two charges are moved closer together?
  a. The electric field doubles.
  b. Coulomb's law takes effect.
  c. The total charge increases.
  d. The force between the charges increases.

_____**15.** Resultant force on a charge is the _____ sum of individual forces on that charge.
  a. scalar
  b. vector
  c. individual
  d. negative

_____**16.** The electric field just outside a charged conductor in electrostatic
  equilibrium is
  a. zero.
  b. at its minimum level.
  c. the same as it is in the center of the conductor.
  d. perpendicular to the conductor's surface.

_____**17.** For a conductor that is in electrostatic equilibrium, any excess charge
  a. flows to the ground.
  b. resides entirely on the conductor's outer surface.
  c. resides entirely on the conductor's interior.
  d. resides entirely in the center of the conductor.

_____**18.** If an irregularly shaped conductor is in electrostatic equilibrium,
  charge accumulates
  a. where the radius of curvature is smallest.
  b. where the radius of curvature is largest.
  c. evenly throughout the conductor.
  d. in flat places.

SHORT ANSWER

19. Materials, such as glass, in which electric charges do not move freely are
  called electrical _____ .

_____

20. Any force between two objects that are not touching is called a (n)
  _____ force.

_____

21. Draw the lines of force representing the electric field surrounding two objects
  that have equal magnitude charges of opposite polarity.

22. The space around a charged object contains an electric _____.

_____

PROBLEM

23. What is the electric force between an electron and a proton that are separated by a distance of 1.0×10^{-10} m? Is the force attractive or repulsive? ($e = 1.60 \times 10^{-19}$ C, $k_C = 8.99 \times 10^9$ N•m^2/C^2)

24. An electron is separated from a potassium nucleus (charge $19e$) by a distance of 5.2×10^{-10} m. What is the electric force between these particles? ($e = 1.60 \times 10^{-19}$ C, $k_C = 8.99 \times 10^9$ N•m^2/C^2)

25. Charge A and charge B are 2.2 m apart. Charge A is 1.0 C, and charge B is 2.0 C. Charge C, which is 2.0 C, is located between them and is in electrostatic equilibrium. How far from charge A is charge C?

Chapter Test B

Electric Forces and Fields
MULTIPLE CHOICE

In the space provided, write the letter of the term or phrase that best completes each statement or best answers each question.

_____ **1.** If a positively charged glass rod is used to charge a metal bar by induction, the charge on the bar
  a. will be equal in magnitude to the charge on the glass rod.
  b. must be negative.
  c. must be positive.
  d. will be greater in magnitude than the charge on the glass rod.

_____ **2.** Which sentence best describes electrical conductors?
  a. Electrical conductors have low mass density.
  b. Electrical conductors have high tensile strength.
  c. Electrical conductors have electric charges that move freely.
  d. Electrical conductors are poor heat conductors.

_____ **3.** Which statement is the most correct regarding electric insulators?
  a. Charges within electric insulators do not readily move.
  b. Electric insulators have high tensile strength.
  c. Electric charges move freely in electric insulators.
  d. Electric insulators are good heat conductors.

_____ **4.** When a charged body is brought close to an uncharged body without touching it, a(n) _____ charge may result on the uncharged body. When a charged body is brought into contact with an uncharged body and then is removed, a(n) _____ charge may result on the uncharged body.
  a. negative; positive   **c.** induced; residual
  b. positive; negative   **d.** residual; induced

_____ **5.** Two point charges, initially 2 cm apart, are moved to a distance of 10 cm apart. By what factor does the resulting electric force between them change?
  a. 25   **c.** $\frac{1}{5}$
  b. 5   **d.** $\frac{1}{25}$

_____ **6.** If the charge is tripled for two identical charges maintained at a constant separation, the electric force between them will be changed by what factor?
  a. $\frac{1}{9}$ **c.** 9
  b. $\frac{2}{3}$ **d.** 18

_____ **7.** Two point charges, initially 1 cm apart, are moved to a distance of 3 cm apart. By what factor do the resulting electric and gravitational forces between them change?

    a. 3   **c.** $\frac{1}{3}$

    b. 9   **d.** $\frac{1}{9}$

_____ **8.** Two positive charges, each of magnitude q, are on the y-axis at points $y = +a$ and $y = -a$. Where would a third positive charge of the same magnitude be located for the net force on the third charge to be zero?

    a. at the origin   **c.** at $y = -2a$

    b. at $y = 2a$ **d.** at $y = -a$

_____ **9.** Which is the most correct statement regarding the drawing of electric field lines?

    a. Electric field lines always connect from one charge to another.

    b. Electric field lines always form closed loops.

    c. Electric field lines can start on a charge of either polarity.

    d. Electric field lines never cross each other.

_____ **10.** If an irregularly shaped conductor is in electrostatic equilibrium, charge accumulates

    a. where the radius of curvature is smallest.

    b. where the radius of curvature is largest.

    c. evenly throughout the conductor.

    d. in flat places.

SHORT ANSWER

11. Explain what happens when you vigorously rub your wool socks on a carpeted floor, touch a metal doorknob, and get a shock.

_____

_____

_____

_____

12. What property was discovered in Millikan's experiments? Explain this property.

_____

_____

_____

_____

13. Describe how a nonconducting material, such as paper, becomes attracted to a negatively charged object brought near it.

_____

_____

_____

_____

14. A negatively charged rubber rod is brought near a neutral, conductive sphere that has no charge. As a result, the part of the sphere closest to the rod becomes positively charged. Explain how this positive charge occurs.

_____

_____

_____

15. In the figure shown on the right, why do only half of the lines originating from the positive charge terminate on the negative charge?

_____

_____

_____

_____

PROBLEM

16. Two equal charges are separated by 3.7×10^{-10} m. The force between the charges has a magnitude of 2.37×10^{-3} N. What is the magnitude of q on the charges? ($k_C = 8.99 \times 10^9$ N•m^2/C^2)

17. An alpha particle (charge $2e$) is sent at high speed toward a gold nucleus. The electric force acting on the alpha particle is 91.0 N when it is 2.00×10^{-14} m away from the gold nucleus. What is the charge on the gold nucleus, as a whole number multiple of e? ($e = 1.60 \times 10^{-19}$ C, $k_C = 8.99 \times 10^9$ N•m²/C²)

18. Two charges are located on the positive x-axis of a coordinate system. Charge $q_1 = 2.00 \times 10^{-9}$C, and it is 0.020 m from the origin. Charge $q_2 = -3.00 \times 10^{-9}$C, and it is 0.040 m from the origin. What is the electric force exerted by these two charges on a third charge, $q_3 = 5.00 \times 10^{-9}$, located at the origin? ($k_C = 8.99 \times 10^9$ N•m²/C²)

19. Two point charges are 4.0 cm apart and have values of 30.0 μC and -30.0 μC, respectively. What is the electric field at the midpoint between the two charges? ($k_C = 8.99 \times 10^9$ N•m²/C²)

20. Charges of 4.0 μC and -6.0 μC are placed at two corners of an equilateral triangle with sides of 0.10 m. What is the magnitude of the electric field created by these two charges at the third corner of the triangle?

Chapter Test A

Electrical Energy and Current
MULTIPLE CHOICE

In the space provided, write the letter of the term or phrase that best completes each statement or best answers each question.

_____ 1. Which of the following is *not* a characteristic of electrical potential energy?
   a. It is a form of mechanical energy.
   b. It results from a single charge.
   c. It results from the interaction between charges.
   d. It is associated with a charge in an electric field.

_____ 2. Two positive point charges are initially separated by a distance of 2 cm. If their separation is increased to 6 cm, the resultant electrical potential energy is equal to what factor multiplied by the initial electrical potential energy?
   a. 3
   b. 9
   c. $\dfrac{1}{3}$
   d. $\dfrac{1}{9}$

_____ 3. Charge transfer between the plates of a capacitor stops when
   a. there is no net charge on the plates.
   b. unequal amounts of charge accumulate on the plates.
   c. the potential difference between the plates is equal to the applied potential difference.
   d. the charge on both plates is the same.

_____ 4. When a capacitor discharges,
   a. it must be attached to a battery.
   b. charges move through the circuit from one plate to the other until both plates are uncharged.
   c. charges move from one plate to the other until equal and opposite charges accumulate on the two plates.
   d. it cannot be connected to a material that conducts.

_____ 5. A parallel-plate capacitor has a capacitance of C F. If the area of the plates is doubled while the distance between the plates is halved, the new capacitance will be
   a. $2\,C$.
   b. $4\,C$.
   c. $\dfrac{C}{2}$.
   d. $\dfrac{C}{4}$.

| Chapter Test A *continued*

_____ **6.** A 1.5 μF capacitor is connected to a 9.0 V battery. Use the expression
$PE = \frac{1}{2}C(\Delta V)^2$ to determine how much energy is stored in the capacitor.
  a. 1.1×10^{-11} J   **c.** 6.1×10^{-2} J
  b. 6.1×10^{-5} J   **d.** 60.8

_____ **7.** How is current affected if the number of charge carriers decreases?
  a. The current increases.
  b. The current decreases.
  c. The current initially decreases and then is gradually restored.
  d. The current is not affected.

_____ **8.** A flashlight bulb with a potential difference of 4.5 V across it has a
resistance of 8.0 Ω. How much current is in the bulb filament?
  a. 36 A   **c.** 1.8 A
  b. 9.4 A   **d.** 0.56 A

_____ **9.** When electrons move through a metal conductor,
  a. they move in a straight line through the conductor.
  b. they move in zigzag patterns because of repeated collisions with the
vibrating metal atoms.
  c. the temperature of the conductor decreases.
  d. they move at the speed of light in a vacuum.

_____ **10.** What is the potential difference across a 5.0 Ω resistor that carries a
current of 5.0 A?
  a. 1.0×10^2 V   **c.** 10.0 V
  b. 25 V   **d.** 1.0 V

_____ **11.** Which of the following does *not* affect a material's resistance?
  a. the length of the material   **c.** the temperature of the material
  b. the type of material   **d.** Ohm's law

_____ **12.** Which of the following wires would have the *least* resistance, assuming that all of the wires have the same cross-sectional area?
  a. an iron wire 10 cm in length   **c.** a copper wire 10 cm in length
  b. an iron wire 5 cm in length   **d.** a copper wire 5 cm in length

_____ **13.** The power ratings on lightbulbs are measures of the
  a. rate that they give off heat and light.
  b. voltage they require.
  c. density of the charge carriers.
  d. amount of negative charge passing through them.

_____ **14.** If a 75 W lightbulb operates at a voltage of 120 V, what is the current in
the bulb?
  a. 0.62 A   **c.** 1.95×10^2 A
  b. 1.6 A   **d.** 9.0×10^3 A

_____**15.** Tripling the current in a circuit with constant resistance has the effect of changing the power by what factor?

   a. $\dfrac{1}{3}$ **c.** 3

   b. $\dfrac{1}{9}$ **d.** 9

SHORT ANSWER

16. What is the source of the energy produced by a battery?

_____

_____

17. Explain why there is a limit to the amount of charge that can be stored in a capacitor.

_____

_____

18. What is the relationship between the radius of a sphere and the capacitance of the sphere?

_____

_____

19. In a circuit connected to a 60 Hz power line, a wire carries a current of 5 A. How far will an electron move through the wire in 60 min? Explain.

_____

_____

20. With regard to the flow of electric current, what is resistance?

_____

_____

21. Which type of electric current is supplied to homes and businesses? Why?

_____

_____

_____

22. What is electric power?

_____

_____

PROBLEM

23. What is the electric potential at a distance of 0.15 m from a point charge of 6.0 μC? ($k_C = 8.99 \times 10^9$ N•m²/C²)

24. A 0.47 μF capacitor holds 1.0 μC of charge on each plate. What is the potential difference across the capacitor?

25. A bolt of lightning discharges 9.7 C in 8.9×10^{-5} s. What is the average current during the discharge?

Assessment

Chapter Test B

Electrical Energy and Current
MULTIPLE CHOICE

In the space provided, write the letter of the term or phrase that best completes each statement or best answers each question.

_____ 1. When a positive charge moves in the direction of the electric field, what happens to the electrical potential energy associated with the charge?
   a. It increases.
   b. It decreases.
   c. It remains the same.
   d. It sharply increases, and then decreases.

_____ 2. When comparing the net charge of a charged capacitor with the net charge of the same capacitor when it is uncharged, the net charge is
   a. greater in the charged capacitor.
   b. less in the charged capacitor.
   c. equal in both capacitors.
   d. greater or less in the charged capacitor, but never equal.

_____ 3. A 0.25 μF capacitor is connected to a 9.0 V battery. What is the charge on the capacitor?
   a. 1.2×10^{-12} C   c. 2.5×10^{-6} C
   b. 2.2×10^{-6} C   d. 2.8×10^{-2} C

_____ 4. A 0.50 μF capacitor is connected to a 12 V battery. How much electrical potential energy is stored in the capacitor?
   a. 3.0×10^{-6} J   c. 1.0×10^{-5} J
   b. 6.0×10^{-6} J   d. 3.6×10^{-5} J

_____ 5. The current in an electron beam in a cathode-ray tube is 7.0×10^{-5} A. How much charge hits the screen in 5.0 s?
   a. 2.8×10^3 C   c. 3.5×10^{-4} C
   b. 5.6×10^{-2} C   d. 5.3×10^{-6} C

_____ 6. When you flip a switch to turn on a light, the delay time before the light turns on is determined by
   a. the number of electron collisions per second in the wire.
   b. the drift speed of the electrons in the wire.
   c. the speed of the electric field moving in the wire.
   d. the resistance of the wire.

_____ **7.** A flashlight bulb with a potential difference of 4.5 V across it has a resistance of 8.0 Ω. How much current is in the bulb filament?

    a. 36 A   **c.** 1.8 A

    b. 9.4 A   **d.** 0.56 A

_____ **8.** Which of the following wires would have the *greatest* resistance?

    a. an aluminum wire 10 cm in length and 3 cm in diameter

    b. an aluminum wire 5 cm in length and 3 cm in diameter

    c. an aluminum wire 10 cm in length and 5 cm in diameter

    d. an aluminum wire 5 cm in length and 5 cm in diameter

_____ **9.** Which set of information will allow you to calculate the kilowatt•hr usage?

    a. the voltage and current in the circuit

    b. the resistance, the current, and the time the circuit operates

    c. the voltage and the resistance of the circuit

    d. the current and the time the circuit operates

_____ **10.** If the current through a 5.00×10^2 W heater is 4.00 A, what is the potential difference across the ends of the heating element?

    a. 2.00×10^3 V   **c.** 2.50×10^1 V

    b. 1.25×10^2 V   **d.** 8.00×10^{-3} V

_____ **11.** If a 325 W heater has a current of 6.0 A, what is the resistance of the heating element?

    a. 88 Ω   **c.** 9.0 Ω

    b. 54 Ω   **d.** 4.5 Ω

_____ **12.** How much does it cost to operate a 695 W heater for exactly 30.0 min if electrical energy costs $0.060 per kW•h?

    a. $0.02   **c.** $0.18

    b. $0.90   **d.** $0.36

SHORT ANSWER

13. What is the result when the energy produced by a battery moves charges from one terminal to the other?

_____

_____

14. Which is the most effective way to increase the amount of energy stored in a capacitor—increasing the capacitance of the capacitor or increasing the voltage across the capacitor? Explain.

_____

_____

_____

15. For a given plate spacing, a capacitor with a dielectric can operate at a higher voltage than it can without the dielectric. Why?

_____

_____

_____

16. An operating (turned on) light bulb has a resistance that is about 10 to 20 times greater than when it is not operating (turned off). How does this explain why light bulbs often burn out at the moment they are turned on?

_____

_____

_____

17. The operating voltage of a power line is increased from 300 kV to 600 kV. If the amount of power delivered to customers over the power line remains constant, how does the power loss in the power line at 300 kV compare to the power loss at 600 kV? Explain.

_____

_____

_____

PROBLEM

18. At what distance from a point charge of 8.0 μC would the electric potential be 4.2×10^4 V? ($k_C = 8.99 \times 10^9$ N•m^2/C^2)

19. A 3.2 μF capacitor has a potential difference of 21.0 V between its plates. How much additional charge flows into the capacitor if the potential difference is increased to 47.0 V?

20. A toaster is connected across a 123 V outlet and dissipates 0.95 kW in the form of electromagnetic radiation and heat. What is the resistance of the toaster?

Chapter Test A

Circuits and Circuit Elements
MULTIPLE CHOICE

In the space provided, write the letter of the term or phrase that best completes each statement or best answers each question.

_____ **1.** Which of the following is the best description of a schematic diagram?
  a. uses pictures to represent the parts of a circuit
  b. determines the location of the parts of a circuit
  c. shows the parts of a circuit and how the parts connect to each other
  d. shows some of the parts that make up a circuit

_____ **2.** A circuit has a continuous path through which charge can flow from a voltage source to a device that uses electrical energy. What is the name of this type of circuit?
  a. a short circuit   **c.** an open circuit
  b. a closed circuit   **d.** a circuit schematic

_____ **3.** How does the potential difference across the bulb in a flashlight compare with the terminal voltage of the batteries used to power the flashlight?
  a. The potential difference is greater than the terminal voltage.
  b. The potential difference is less than the terminal voltage.
  c. The potential difference is equal to the terminal voltage.
  d. It cannot be determined unless the internal resistance of the batteries is known.

_____ **4.** Three resistors connected in series carry currents labeled I_1, I_2, and I_3, respectively. Which of the following expresses the total current, I_t, in the system made up of the three resistors in series?
  a. $I_t = I_1 + I_2 + I_3$ **c.** $I_t = I_1 = I_2 = I_3$
  b. $I_t = \left(\dfrac{1}{I_1} + \dfrac{1}{I_2} + \dfrac{1}{I_3} \right)$ **d.** $I_t = \left(\dfrac{1}{I_1} + \dfrac{1}{I_2} + \dfrac{1}{I_3} \right)^{-1}$

_____ **5.** Three resistors connected in series have potential differences across them labeled ΔV_1, ΔV_2, and ΔV_3. Which of the following expresses the potential difference taken over the three resistors together?
  a. $\Delta V_t = \Delta V_1 + \Delta V_2 + \Delta V_3$
  b. $\Delta V_t = \left(\dfrac{1}{\Delta V_1} + \dfrac{1}{\Delta V_2} + \dfrac{1}{\Delta V_3} \right)$
  c. $\Delta V_t = \Delta V_1 = \Delta V_2 = \Delta V_3$
  d. $\Delta V_t = \left(\dfrac{1}{\Delta V_1} + \dfrac{1}{\Delta V_2} + \dfrac{1}{\Delta V_3} \right)^{-1}$

| Chapter Test A *continued*

_____ **6.** Three resistors with values of R_1, R_2, and R_3 are connected in series. Which of the following expresses the total resistance, R_{eq}, of the three resistors?

    a. $R_{eq} = R_1 + R_2 + R_3$

    c. $R_{eq} = R_1 = R_2 = R_3$

    b. $R_{eq} = \left(\dfrac{1}{R_1} + \dfrac{1}{R_2} + \dfrac{1}{R_3} \right)$

    d. $R_{eq} = \left(\dfrac{1}{R_1} + \dfrac{1}{R_2} + \dfrac{1}{R_3} \right)^{-1}$

_____ **7.** Three resistors connected in parallel carry currents labeled I_1, I_2, and I_3. Which of the following expresses the total current I_t in the combined system?

    a. $I_t = I_1 + I_2 + I_3$

    c. $I_t = I_1 = I_2 = I_3$

    b. $I_t = \left(\dfrac{1}{I_1} + \dfrac{1}{I_2} + \dfrac{1}{I_3} \right)$

    d. $I_t = \left(\dfrac{1}{I_1} + \dfrac{1}{I_2} + \dfrac{1}{I_3} \right)^{-1}$

_____ **8.** Three resistors connected in parallel have potential differences across them labeled ΔV_1, ΔV_2, and ΔV_3. Which of the following expresses the potential difference across all three resistors?

    a. $\Delta V_t = \Delta V_1 + \Delta V_2 + \Delta V_3$

    c. $\Delta V_t = \Delta V_1 = \Delta V_2 = \Delta V_3$

    b. $\Delta V_t = \left(\dfrac{1}{\Delta V_1} + \dfrac{1}{\Delta V_2} + \dfrac{1}{\Delta V_3} \right)$

    d. $\Delta V_t = \left(\dfrac{1}{\Delta V_1} + \dfrac{1}{\Delta V_2} + \dfrac{1}{\Delta V_3} \right)^{-1}$

_____ **9.** Three resistors with values of R_1, R_2, and R_3 are connected in parallel. Which of the following expresses the total resistance, R_{eq}, of the three resistors?

    a. $R_{eq} = R_1 + R_2 + R_3$

    c. $R_{eq} = R_1 = R_2 = R_3$

    b. $R_{eq} = \left(\dfrac{1}{R_1} + \dfrac{1}{R_2} + \dfrac{1}{R_3} \right)$

    d. $R_{eq} = \left(\dfrac{1}{R_4} + \dfrac{1}{R_2} + \dfrac{1}{R_3} \right)^{-1}$

_____ **10.** Three resistors with values of 3.0 Ω, 6.0 Ω, and 12 Ω are connected in parallel. What is the equivalent resistance of this combination?

    a. 0.26 Ω

    c. 9.0 Ω

    b. 1.7 Ω

    d. 33 Ω

_____ **11.** The equivalent resistance of a complex circuit is usually determined by

    a. inspection.

    b. simplifying the circuit into groups of series and parallel circuits.

    c. adding and subtracting individual resistances.

    d. dividing the sum of the individual resistances by the number of resistances.

_____ **12.** To find the current in a complex circuit, it is necessary to know the

    a. potential difference in each device in the circuit.

    b. current in each device in the circuit.

    c. equivalent resistance of the circuit.

    d. number of branches in the circuit.

_____13. Two resistors with values of 6.0 Ω and 12 Ω are connected in parallel. This combination is connected in series with a 4.0 Ω resistor. What is the equivalent resistance of this combination?
   a. 0.50 Ω   **c.** 8.0 Ω
   b. 2.0 Ω   **d.** 22 Ω

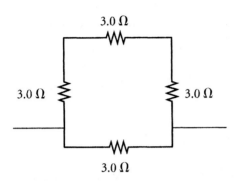

_____14. What is the equivalent resistance for the resistors in the figure shown above?
   a. 1.3 Ω   **c.** 3.0 Ω
   b. 2.2 Ω   **d.** 0.75 Ω

_____15. In any complex resistance circuit, the voltage across any resistor in the circuit is always
   a. less than the voltage source.
   b. equal to or less than the voltage source.
   c. equal to the voltage source.
   d. greater than the voltage source.

SHORT ANSWER

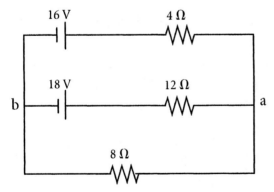

16. Identify the types of elements in the schematic diagram above and the number of each type.

_____

_____

17. Which bulb or bulbs will have a current in the schematic diagram shown on the right?

_____

_____

18. Why does the potential difference measured at the terminals of a battery decrease as the amount of current supplied to a load increases?

_____

_____

PROBLEM

19. A current of 0.20 A passes through a 3.0 Ω resistor. The resistor is connected in series with a 6.0 V battery and an unknown resistor. What is the resistance value of the unknown resistor?

20. Three resistors with values of 27 Ω, 81 Ω, 16 Ω, respectively, are connected in parallel. What is their equivalent resistance?

Chapter Test B

Circuits and Circuit Elements
MULTIPLE CHOICE

In the space provided, write the letter of the term or phrase that best completes each statement or best answers each question.

_____ **1.** What happens when the switch is closed in the circuit shown on the right?
  a. The lamp lights because current from the battery flows through the lamp.
  b. Current from the battery flows through the resistor.
  c. Current from the battery flows through both the lamp and the resistor.
  d. The lamp goes out, because the battery terminals connect to each other.

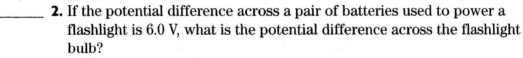

_____ **2.** If the potential difference across a pair of batteries used to power a flashlight is 6.0 V, what is the potential difference across the flashlight bulb?
  a. 3.0 V   **c.** 9.0 V
  b. 6.0 V   **d.** 12 V

_____ **3.** Which of the following statements about a battery as a source of electric current is *not* true?
  a. A battery is a source of emf.
  b. A battery provides the energy that moves charge.
  c. The terminal voltage of a battery is equal to its emf.
  d. The terminal voltage of a battery is the voltage it delivers to the load.

_____ **4.** Three resistors with values of 4.0 Ω, 6.0 Ω, and 8.0 Ω, respectively, are connected in series. What is their equivalent resistance?
  a. 18 Ω   **c.** 6.0 Ω
  b. 8.0 Ω   **d.** 1.8 Ω

_____ **5.** A circuit is composed of resistors wired in series. What is the relationship between the equivalent resistance of the circuit and the resistance
of the individual resistors?

   a. The equivalent resistance is equal to the largest resistance in the
   circuit.

   b. The equivalent resistance is greater than the sum of all the resistances in the circuit.

   c. The equivalent resistance is equal to the sum of the individual
   resistances.

   d. The equivalent resistance is less than the smallest resistance in the
   circuit.

_____ **6.** Two resistors having the same resistance value are wired in parallel.
How does the equivalent resistance compare to the resistance value of
a single resistor?

   a. The equivalent resistance is twice the value of a single resistor.

   b. The equivalent resistance is the same as a single resistor.

   c. The equivalent resistance is half the value of a single resistor.

   d. The equivalent resistance is greater than that of a single resistor.

_____ **7.** Three resistors with values of 4.0 Ω, 6.0 Ω, and 10.0 Ω are connected
in parallel. What is their equivalent resistance?

   a. 20.0 Ω

   b. 7.3 Ω

   c. 6.0 Ω

   d. 1.9 Ω

_____ **8.** What is the equivalent resistance of the
resistors in the figure shown on the right?

   a. 7.5 Ω

   b. 10 Ω

   c. 16 Ω

   d. 18 Ω

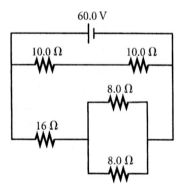

_____ **9.** What is the equivalent resistance for the
resistors in the figure shown on the right?

   a. 25 Ω

   b. 10.0 Ω

   c. 7.5 Ω

   d. 5.0 Ω

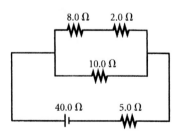

| Chapter Test B *continued*

_____ **10.** Three resistors connected in parallel
          have individual values of 4.0 Ω, 6.0 Ω,
          and 10.0 Ω, as shown on the right.
          If this combination is connected in
          series with a 12.0 V battery and a
          2.0 Ω resistor, what is the current
          in the 10.0 Ω resistor?

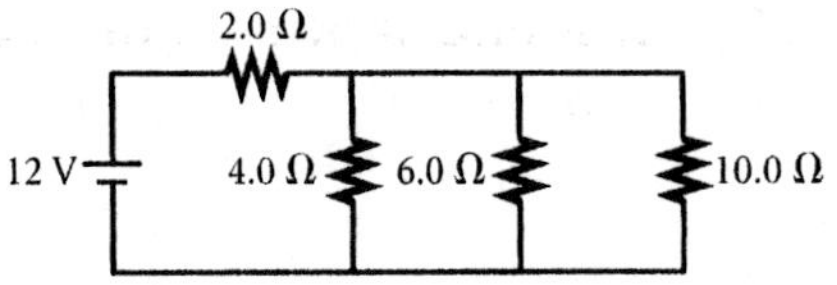

        a. 0.58 A   **c.** 11 A

        b. 1.0 A   **d.** 16 A

SHORT ANSWER

11. Draw a schematic diagram that contains a 1000 V battery, a 3000 Ω resistor, a
0.5 μF capacitor, and an open switch wired as a series circuit.

12. What part of a circuit dissipates energy?

_____

13. In the circuit shown on the right, what will happen when the
switch is closed? Explain.

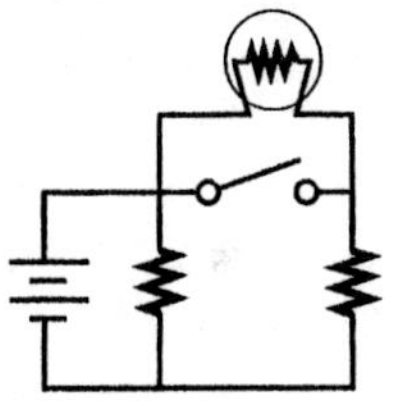

_____

_____

14. In a series circuit, one of the resistors is replaced with a resistor having a
lower resistance value. How does this affect the current in the circuit?
Explain.

_____

_____

15. You have seven resistors available, and all of the resistors have a value of
100.0 Ω. How would you connect these seven resistors to produce an equivalent resistance of 70.0 Ω?

_____

_____

_____

| Chapter Test B *continued*

16. In the circuit shown on the right, which resistors, if any, have equal voltages across them?

Ra · Rb · Rc

Rd · Re · Rf

6.0 V

_____

_____

_____

17. A voltmeter is used to measure the voltage across a device and is placed in parallel with the device. All voltmeters have resistance. Why would a voltmeter with a high resistance be preferable to one with low resistance when measuring the voltage across a resistor that is part of a complex circuit?

_____

_____

_____

PROBLEM

18. A current of 0.20 A passes through a 3.0 Ω resistor. The resistor is connected in series with a 6.0 V battery and an unknown resistor. What is the resistance value of the unknown resistor?

19. Three resistors with values of 15 Ω, 41 Ω, 58 Ω, respectively, are connected in parallel. What is their equivalent resistance?

20. How much current is in one of the 10 Ω resistors in the diagram shown on the right?

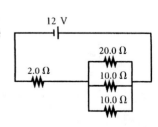

12 V

20.0 Ω

2.0 Ω

10.0 Ω

10.0 Ω

Chapter Test A

Magnetism

MULTIPLE CHOICE

In the space provided, write the letter of the term or phrase that best completes each statement or best answers each question.

_____ **1.** Which of the following situations is *not* true for magnets?
  a. Like poles repel each other.
  b. Unlike poles repel each other.
  c. North poles repel each other.
  d. A north pole and a south pole will attract each other.

_____ **2.** Where is the magnitude of the magnetic field around a permanent magnet greatest?
  a. close to the poles
  b. far from the poles
  c. The magnitude is equal at all points on the field.
  d. The magnitude depends on the material of the magnet.

_____ **3.** One useful way to model magnetic field strength is to define a quantity called magnetic flux Φ_M. Which of the following definitions for magnetic flux, Φ_M, is correct?
  a. the number of field lines that cross a certain area
  b. $AB\cos\theta$
  c. (surface area) $\times$ (magnetic field component normal to the plane of surface)
  d. all of the above

_____ **4.** All of the following statements about magnetic field lines around a permanent magnet are true *except* which one?
  a. Magnetic field lines appear to end at the north pole of a magnet.
  b. Magnetic field lines have no beginning or end.
  c. Magnetic field lines always form a closed loop.
  d. In a permanent magnet, the field lines actually continue within the magnet itself.

_____ **5.** In a magnetized substance, the domains
  a. are randomly oriented.
  b. cancel each other.
  c. line up mainly in one direction.
  d. can never be reoriented.

_____ **6.** In a permanent magnet,
   a. domain alignment persists after the external magnetic field is removed.
   b. domain alignment becomes random after the external magnetic field is removed.
   c. domains are always randomly oriented.
   d. the magnetic fields of the domains cancel each other.

_____ **7.** In soft magnetic materials such as iron, what happens when an external magnetic field is removed?
   a. The domain alignment persists.
   b. The orientation of domains fluctuates.
   c. The material becomes a hard magnetic material.
   d. The material returns to an unmagnetized state.

_____ **8.** Which statement describes Earth's magnetic declination?
   a. the angle between Earth's magnetic field and Earth's surface
   b. Earth's magnetic field strength at the equator
   c. the tendency for Earth's field to reverse itself
   d. the angle between true north and north indicated by a compass

_____ **9.** According to the right-hand rule, if a current-carrying wire is grasped in the right hand with the thumb in the direction of the current, the four fingers will curl in the direction of
   a. the magnetic force, $F_{magnetic}$.
   b. the magnetic field, B.
   c. the current's velocity, v.
   d. the current's path, P.

_____ **10.** The lines of the magnetic field around a current-carrying wire
   a. point away from the wire.
   b. point toward the wire.
   c. form concentric circles around the wire.
   d. are parallel with the wire.

_____ **11.** The direction of the force on a current-carrying wire in an external magnetic field is
   a. perpendicular to the current only.
   b. perpendicular to the magnetic field only.
   c. perpendicular to both the current and the magnetic field.
   d. parallel to the current and to the magnetic field.

_____**12.** What is the path of an electron moving perpendicular to a uniform
magnetic field?
  a. straight line
  b. circle
  c. ellipse
  d. parabola

_____**13.** What is the path of an electron moving parallel to a uniform
magnetic field?
  a. straight line
  b. circle
  c. ellipse
  d. parabola

_____**14.** A stationary positive charge, Q, is located in a magnetic field, B, which
is directed toward the right. The direction of the magnetic force on Q is
  a. toward the right.
  b. up.
  c. down.
  d. There is no magnetic force.

_____**15.** Consider two long, straight, parallel wires, each carrying a current I.
If the currents move in opposite directions,
  a. the two wires will attract each other.
  b. the two wires will repel each other.
  c. the two wires will exert a torque on each other.
  d. neither wire will exert a force on the other.

_____**16.** Consider two long, straight, parallel wires, each carrying a current I.
If the currents move in the same direction,
  a. the two wires will attract each other.
  b. the two wires will repel each other.
  c. the two wires will exert a torque on each other.
  d. neither wire will exert a force on the other.

SHORT ANSWER

17. Why do magnetic poles always occur in pairs?

_____

_____

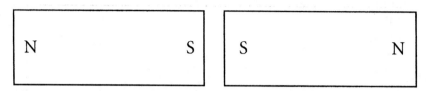

18. Will the magnets in the figure above attract or repel each other?

_____

_____

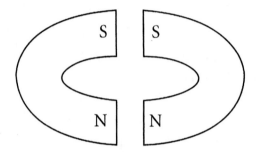

19. Will the magnets in the figure above attract or repel each other?

_____

_____

PROBLEM

20. An electron moves north at a velocity of 9.8×10^4 m/s and has a magnetic force of 5.6×10^{-18} N west exerted on it. If the magnetic field points upward, what is the magnitude of the magnetic field?

Assessment

Chapter Test B

Magnetism
MULTIPLE CHOICE

In the space provided, write the letter of the term or phrase that best completes each statement or best answers each question.

_____ **1.** Which compass needle orientation in the figure on the right might correctly describe the magnet's field at that point?
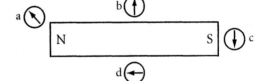
   a. a
   b. b
   c. c
   d. d

_____ **2.** Which of the following statements about Earth's magnetic field is true?
   a. The geographic North Pole of Earth and Earth's magnetic north pole are at the same location.
   b. The geographic South Pole of Earth and Earth's magnetic north pole are relatively close to each other.
   c. The north needle of a compass always points to the geographic North Pole of Earth.
   d. The north needle of a compass points to Earth's magnetic north pole.

_____ **3.** Which of the following modifications to a solenoid would be most likely to decrease the strength of its magnetic field?
   a. removing its iron rod core and increasing the number of coils
   b. increasing the current and reducing the number of coils
   c. reducing the number of coils and inserting an iron core
   d. decreasing the current and reducing the number of coils

_____ **4.** Under which of the following conditions is the net magnetic force on a charged particle equal to zero?
   a. when the particle is stationary
   b. when the particle is moving parallel to the magnetic field
   c. when the particle is not charged
   d. all of the above

_____ **5.** An electron moves north at a velocity of 4.5×10^4 m/s and has a force of 7.2×10^{-18} N exerted on it. If the magnetic field points upward, what is the magnitude of the magnetic field?
   a. 1.0 mT   **c.** 3.6 mT
   b. 2.0 mT   **d.** 4.8 mT

_____ **6.** An electron that moves with a speed of 3.0×10^4 m/s perpendicular to a uniform magnetic field of 0.40 T experiences a force of what magnitude? ($q_e = 1.60 \times 10^{-19}$ C)

    a. 2.2×10^{24} N   **c.** 4.8×10^{-14} N

    b. 1.9×10^{-15} N   **d.** 0 N

_____ **7.** An electron moves across Earth's equator at a speed of 2.5×10^6 m/s and in a direction 35° north of east. At this point, Earth's magnetic field has a direction due north, is parallel to the surface, and has a value of 0.10×10^{-4} T. What is the magnitude of the force acting on the electron due to its interaction with Earth's magnetic field? ($q_e = 1.60 \times 10^{-19}$ C)

    a. 5.1×10^{-18} N   **c.** 3.3×10^{-18} N

    b. 4.0×10^{-18} N   **d.** 2.3×10^{-18} N

_____ **8.** If a proton is released at the equator and falls toward Earth under the influence of gravity, the magnetic force on the proton will be toward the _____ assuming the magnetic field is directed toward the north at this location.

    a. north   **c.** east

    b. south   **d.** west

_____ **9.** A 2.0 m wire segment carrying a current of 0.60 A oriented parallel to a uniform magnetic field of 0.50 T experiences a force of what magnitude?

    a. 0.60 N   **c.** 0.15 N

    b. 0.30 N   **d.** 0.0 N

_____ **10.** A current-carrying conductor in and perpendicular to a magnetic field experiences a force that is

    a. perpendicular to the current.

    b. parallel to the current.

    c. inversely proportional to the potential difference.

    d. inversely proportional to the velocity.

SHORT ANSWER

11. A bar magnet is suspended and allowed to rotate freely. If the magnetic field of Earth is considered to be equivalent to that of a large bar magnet, which pole of the suspended magnet would point toward the magnetic north pole of Earth?

_____

_____

12. If the head of an iron nail touches a magnet, the nail will become a magnet by induction. If the nail touches the north pole of the magnet, what kind of pole is at the point of the nail? Explain.

_____

_____

_____

_____

13. Use the magnetic domain model to compare and contrast hard and soft magnetic materials.

_____

_____

_____

_____

_____

_____

14. The magnetic field of a bar magnet is shown in the figure to the right. Is the magnet's north pole at A or B?

_____

_____

15. Use the right-hand rule to determine the direction of the magnetic field within the loop from the figure to the right. Since the magnetic field of a current-carrying loop resembles that of a bar magnet, is the north pole of the current-carrying loop above or below the loop?

_____

_____

_____

_____

16. What is the name of the device shown on the right? Which end is the south pole? Is the current entering or leaving the wire coil at the top right?

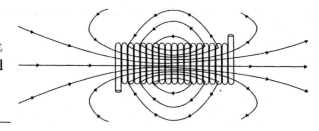

_____

_____

_____

17. Find the direction of the force on a proton moving through the magnetic field shown on the right.

_____

_____

PROBLEM

18. An electron is moving parallel to the Earth's surface at the equator in a direction 25° south of east. Its velocity is 7.3×10^4 m/s and a magnetic force of 1.8×10^{-18} N is exerted on the electron. If the magnetic field points south at this location, what is the direction of the magnetic force on the electron and the magnitude of the magnetic field?

19. A proton moves perpendicularly to a magnetic field that has a magnitude of 6.48×10^{-2} T. A magnetic force of 7.16×10^{-14} N is acting on it. If the proton moves a total distance of 0.500 m in the magnetic field, how long does it take for the proton to move across the magnetic field? If the magnetic force is directed north and the magnetic field is directed upward, what was the proton's velocity?

20. A wire 48 m long carries a current of 18 A from west to east. If a magnetic field of 8.3×10^{-4} T directed toward the south is acting on the wire, find the direction and magnitude of the magnetic force.

Chapter Test A

Electromagnetic Induction
MULTIPLE CHOICE

In the space provided, write the letter of the term or phrase that best completes each statement or best answers each question.

_____ 1. A current can be induced in a closed circuit without the use of a battery or an electrical power supply by moving the circuit through a
   a. high temperature field.   **c.** magnetic field.
   b. gravitational field.   **d.** nuclear field.

_____ 2. A loop of wire is held in a vertical position at the equator with the face of the loop facing in the east-west direction. What change will induce the greatest current in the loop?
   a. raising the loop to a higher elevation
   b. moving the loop north
   c. rotating the loop so its face is vertical
   d. rotating the loop so its face is north-south

_____ 3. Electricity may be generated by rotating a loop of wire between the poles of a magnet. The induced current is greatest when
   a. the plane of the loop is parallel to the magnetic field.
   b. the plane of the loop is perpendicular to the magnetic field.
   c. the magnetic flux through the loop is a minimum.
   d. the plane of the loop makes an angle of 45° with the magnetic field.

_____ 4. According to Lenz's law, the magnetic field of an induced current in a conductor will
   a. enhance the applied field.
   b. heat the conductor.
   c. increase the potential difference.
   d. oppose a change in the applied magnetic field.

_____ 5. According to Lenz's law, if the applied magnetic field changes,
   a. the induced field attempts to keep the total field strength constant.
   b. the induced field attempts to increase the total field strength.
   c. the induced field attempts to decrease the total field strength.
   d. the induced field attempts to oscillate about an equilibrium value.

_____ 6. Which of the following options can be used to generate electricity?
   a. Move the circuit loop into and out of a magnetic field.
   b. Change the magnetic field strength around the circuit loop.
   c. Change the orientation of the circuit loop with respect to the magnetic field.
   d. all of the above

_____ **7.** Which conversion process is the basic function of the electric motor?
   a. mechanical energy to electrical energy
   b. electrical energy to mechanical energy
   c. low voltage to high voltage, or vice versa
   d. alternating current to direct current

_____ **8.** In a two-coil system, the mutual inductance depends on
   a. only the geometrical properties of the coils.
   b. only the orientation of the coils to each other.
   c. both the geometrical properties of the coils and their orientation to each other.
   d. neither the geometrical properties of the coils nor their orientation to each other.

_____ **9.** In a primary-secondary coil combination, which of the following conditions is met in the primary coil when the current in the secondary coil is at its maximum?
   a. The current is maximum in a positive direction.
   b. The current is maximum in a negative direction.
   c. The rate of current change is maximum.
   d. The voltage is maximum in a positive direction.

_____ **10.** What is rms (root-mean-square) current?
   a. the value of alternating current that gives the same heating effect that the corresponding value of direct current does
   b. an important measure of the current in an ac circuit
   c. the amount of direct current that would dissipate the same energy in a resistor as is dissipated by the instantaneous alternating current over a complete cycle
   d. all of the above

_____ **11.** A generator's maximum output is 220 V. Calculate the rms potential difference.
   a. 110 V   **c.** 160 V
   b. 150 V   **d.** 310 V

_____ **12.** The rms current in an ac current is 3.6 A. Find the maximum current.
   a. 5.1 A   **c.** 2.8 A
   b. 4.7 A   **d.** 1.8 A

_____ **13.** A step-up transformer used on a 120 V line has 95 turns on the primary and 2850 turns on the secondary. What is the potential difference across the secondary?
   a. 30 V   **c.** 2400 V
   b. 1800 V   **d.** 3600 V

_____14. A potential difference of 115 V across the primary of a step-down transformer provides a potential difference of 2.3 V across the secondary. What is the ratio of the number of turns of wire on the primary to the number of turns on the secondary?
   a. 1:50
   b. 50:1
   c. 25:1
   d. 1:25

_____15. Electromagnetic waves are _____ electric and magnetic fields.
   a. transverse
   b. constant
   c. oscillating
   d. parallel

_____16. All of the following statements about the electromagnetic force are true *except* which one?
   a. It is one of the four fundamental forces in the universe.
   b. It exerts a force on either charged or uncharged particles.
   c. It obeys the inverse-square law.
   d. It is produced by—and produces—electric and magnetic fields.

_____17. Where is energy stored in electromagnetic waves?
   a. in the wave's moving atoms
   b. in the oscillating electric and magnetic fields
   c. in the electromagnetic force
   d. in the wave's directional vector

_____18. What do radio waves, microwaves, X rays, and gamma rays all have in common?
   a. They are produced in the same way.
   b. They are electromagnetic waves.
   c. They are detected in the same way.
   d. They store the same amount of energy.

SHORT ANSWER

19. List three ways to induce a current in a circuit loop, using only a magnet.

_____

_____

_____

20. What is a turbine's rotational motion used for in commercial power plants?

_____

_____

21. The rms emf across a resistor is equal to what?

_____

_____

22. Explain how the terms *energy, electromagnetic force, electromagnetic wave,* and *electromagnetic radiation* are related to one another.

_____

_____

_____

_____

PROBLEM

23. A coil with 275 turns and a cross-sectional area of 0.750 m^2 experiences a magnetic field whose strength increases by +0.900 T in 1.25 s. The plane of the coil moves perpendicularly to the plane of the magnetic field. What is the induced emf in the coil?

24. An ac generator has a maximum output emf of 4.20 $\times$ 10^2 V. What is the rms potential difference?

25. A step-up transformer used on a 120 V line has 38 turns on the primary and 5163 turns on the secondary. What is the potential difference across the secondary?

Chapter Test B

Electromagnetic Induction
MULTIPLE CHOICE

In the space provided, write the letter of the term or phrase that best completes each statement or best answers each question.

_____ **1.** All of the following are ways to induce an emf in a conductor *except* which one?
- **a.** Deflect a charge in a magnetic field.
- **b.** Change the size of the loop in a magnetic field.
- **c.** Connect the conductor to a power supply.
- **d.** Change the strength of the magnetic field.

_____ **2.** A bar magnet falls through a loop of wire with constant velocity, and the north pole enters the loop first. The induced current will be greatest when the magnet is located so that the loop is
- **a.** near either the north or the south pole.
- **b.** near the north pole only.
- **c.** near the middle of the magnet.
- **d.** With no acceleration, the induced current is zero.

_____ **3.** Which statement is correct?
- **a.** The magnetic field of an induced current increases the approaching magnetic field.
- **b.** According to the principle of energy conservation, an induced field attempts to keep the total field strength constant.
- **c.** An induced electric field opposes an applied magnetic field.
- **d.** Lenz's law is used to find the average induced emf.

_____ **4.** A generator with a single loop produces the greatest magnetic force on the charges and the greatest induced emf when
- **a.** the plane of the loop is parallel to the magnetic field.
- **b.** half of the loop segments are moving perpendicular to the magnetic field.
- **c.** the plane of the loop is perpendicular to the magnetic field.
- **d.** none of the above

_____ **5.** Which conversion process is the basic function of the electric generator?
- **a.** mechanical energy to electrical energy
- **b.** electrical energy to mechanical energy
- **c.** low voltage to high voltage, or vice versa
- **d.** alternating current to direct current

_____ **6.** Two loops of wire are arranged so that a changing current in the primary will induce a current in the secondary. The secondary loop has twice as many turns as the primary loop. As long as the current in the primary is steady at 3.0 A, the current in the secondary will be

a. 6.0 A.   **c.** 1.5 A.

b. 3.0 A.   **d.** zero.

_____ **7.** All of the following statements about ac rms values and maximum values are true *except* which one?

a. Rms values may equal maximum values.

b. Rms values are always less than maximum values.

c. Rms values are approximately 70 percent of the maximum values.

d. Rms values are different from maximum values because the alternating current is at its maximum only for an instant.

_____ **8.** An ac generator has a maximum output emf of 215 V. What is the rms potential difference?

a. 145 V   **c.** 216 V

b. 152 V   **d.** 304 V

_____ **9.** A step-down transformer has 2500 turns on its primary and 5.0×10^1 turns on its secondary. If the potential difference across the primary is 4850 V, what is the potential difference across the secondary?

a. 1.0 V   **c.** 110 V

b. 97 V   **d.** 240 V

_____ **10.** How does an electromagnetic wave propagate itself?

a. A changing magnetic field induces an electric field perpendicular to the magnetic field.

b. A changing electric field induces a magnetic field perpendicular to the electric field.

c. Changing electric and magnetic fields produce a transverse wave that is perpendicular to both of the oscillating fields.

d. all of the above

_____ **11.** How many electromagnetic forces exist?

a. one   **c.** three

b. two   **d.** four

_____ **12.** Where is energy stored in electromagnetic waves?

a. in the wave's moving atoms

b. in the oscillating electric and magnetic fields

c. in the electromagnetic force

d. in the wave's directional vector

_____**13.** Which of the following statements about electromagnetic radiation
is true?
    a. It transfers energy to objects in the path of the electromagnetic
waves.
    b. It can be converted to other energy forms.
    c. It transports the energy of electromagnetic waves.
    d. all of the above

SHORT ANSWER

14. Thoroughly describe the magnetic field characteristics as an induced current
is produced in a coil of wire by an approaching and receding magnet.

_____

_____

_____

_____

15. What is back emf? Does it increase or decrease motor efficiency? Explain.

_____

_____

16. How does the term *wave-particle duality of light* relate to electromagnetic
waves? What are these "particles" called? How does the behavior of a high-
energy particle compare to a low-energy particle?

_____

_____

_____

17. Identify one type of electromagnetic wave with a long wavelength. Describe
at least two of its applications, and explain why the selected wave's long
wavelength is important for both applications.

_____

_____

_____

PROBLEM

18. A coil with a wire is wrapped 40.0 turns around a 0.50 m^2 hollow tube. A magnetic field is applied perpendicularly to the plane of the coil. The field changes uniformly from 0.00 T to 0.95 T in 2.0 s. If the resistance in the coil is 3.0 Ω, what is the magnitude of the induced current?

19. The rms potential difference of an ac generator is 6.9 × 10^2 V. What is the maximum output emf?

20. A primary coil with 196 turns and a cross-sectional area of 0.180 m^2 experiences a magnetic field whose strength increases uniformly from 0.000 T to 0.950 T in 0.700 s. The plane of the coil moves perpendicularly to the plane of the magnetic field. If the secondary coil has 9691 turns, what is the maximum induced emf in the secondary coil?

Atomic Physics

MULTIPLE CHOICE

In the space provided, write the letter of the term or phrase that best completes each statement or best answers each question.

_____ **1.** What term is used to describe a perfect radiator and absorber of electromagnetic radiation?
  a. blackbody   **c.** quantum
  b. atom   **d.** photon

_____ **2.** Classical electromagnetic theory predicted that the energy radiated by a blackbody would become infinite as the wavelength became shorter. What was the contradiction between observation and this result called?
  a. the quantum theory   **c.** the wave-particle duality
  b. the photoelectric effect   **d.** the ultraviolet catastrophe

_____ **3.** What were the units of light energy emitted by blackbody radiation originally called?
  a. electron volts   **c.** joules
  b. quanta   **d.** resonators

_____ **4.** According to the Rutherford model, what makes up most of the volume of an atom?
  a. empty space   **c.** positive charges
  b. the nucleus   **d.** electrons

_____ **5.** In Rutherford's experiment, why did the nucleus repel alpha particles?
  a. electrostatic repulsion between the negatively charged nucleus and alpha particles
  b. electrostatic attraction between the negatively charged nucleus and alpha particles
  c. electrostatic repulsion between the positively charged nucleus and alpha particles
  d. electrostatic attraction between the positively charged nucleus and alpha particles

_____ **6.** What is the concentration of positive charge and mass in Rutherford's atomic model called?
  a. alpha particle   **c.** proton
  b. neutron   **d.** nucleus

_____ **7.** Which statement about Rutherford's model of the atom is *not* correct?
  a. The model states that positive charge is unevenly distributed.
  b. The model depicts electrons orbiting the nucleus as planets orbit the sun.
  c. The model explains spectral lines.
  d. The model states that atoms are unstable.

_____ **8.** When a high potential difference is applied to a low-pressure gas, what kind of spectrum will the gas emit?
  a. emission   **c.** continuous
  b. absorption   **d.** monochromatic

_____ **9.** Which statement about emission spectra is correct?
  a. All of the lines are evenly spaced.
  b. All noble gases have the same spectra.
  c. Each line corresponds to a series of wavelengths.
  d. All of the lines result from discrete energy differences.

_____ **10.** What would you observe if light from argon gas were passed through a prism?
  a. a series of discrete bright lines
  b. a continuous spectrum
  c. a series of dark lines imposed on a continuous spectrum
  d. a single bright line

_____ **11.** Which of the following is *not* a feature of Bohr's model of the atom?
  a. Electrons move in circular orbits about the nucleus.
  b. Only certain electron orbits are allowed.
  c. Electrons emit radiation continuously while orbiting the nucleus.
  d. Electron jumps between energy levels account for discrete spectral lines.

_____ **12.** What is the process in which an electron returns to a lower energy level and emits a photon?
  a. spontaneous emission   **c.** line absorption
  b. line emission   **d.** energy transition

_____ **13.** How will light behave in a single experiment, according to the principle of wave-particle duality?
  a. Light will act both like a wave and like a particle.
  b. Light will act either like a wave or like a particle.
  c. Light will act neither like a wave nor like a particle.
  d. Light always exists as two waves or as two particles.

_____14. Which of the following processes is more easily observable for light
with a short wavelength?
a. the photoelectric effect   **c.** diffraction
b. radio transmission   **d.** interference

_____15. According to the Heisenberg uncertainty principle, which of the fol-
lowing statements about the simultaneous measurements of position
and momentum is true?
a. Neither quantity can be measured with accuracy.
b. The more accurately one value is measured, the less accurately the
other value is known.
c. Both quantities can be measured with infinite accuracy.
d. Accuracy of measurement improves as the object observed
becomes less massive.

_____16. What happens as the frequency of photons increases?
a. The diffraction of light becomes easier to observe.
b. The momentum of light decreases.
c. The wave effects of light become easier to observe.
d. The wave effects of light become more difficult to observe.

_____17. What picture of the electron is suggested by the quantum-mechanical
model of the hydrogen atom?
a. a raisin in pudding   **c.** a planetary orbiting body
b. a probability cloud   **d.** a light quantum

_____18. What does the peak of a probability curve for an electron in an atom
indicate?
a. the location where there is zero probability of finding the electron
b. that the electron's location can be precisely determined
c. that Heisenberg's uncertainty principle is violated
d. the distance from the nucleus at which the electron is most likely to
be found

SHORT ANSWER

19. What is an emission spectrum?

_____

_____

_____

20. Which model of light best explains interference phenomena?

_____

21. Which model of light best explains the photoelectric effect?

_____

22. The maximum kinetic energy of the photoelectrons emitted by a metal exposed to light of a given wavelength happens to be equal to the work function of the metal. How does the energy of the incoming photons compare to the maximum kinetic energy of the emitted photoelectrons?

_____

_____

_____

PROBLEM

23. What is the energy, in eV, of a photon whose frequency is 3.0×10^{14} Hz?
($h = 6.63 \times 10^{-34}$ J•s; 1 eV = 1.60×10^{-19} J)

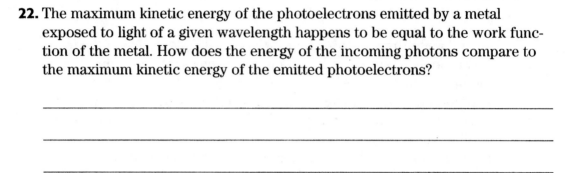

E_6 ———————————————— $E = -0.378$ eV
E_5 ———————————————— $E = -0.544$ eV
E_4 ———————————————— $E = -0.850$ eV

E_3 ———————————————— $E = -1.51$ eV

E_2 ———————————————— $E = -3.40$ eV

24. What is the energy of the photon emitted when the electron in a hydrogen atom drops from energy level E_6 to energy level E_3 in the figure above?

25. What is the de Broglie wavelength for a proton that has a mass of 1.67×10^{-27} kg and is moving at a speed of 1.3×10^3 m/s?
($h = 6.63 \times 10^{-34}$ J•s)

Assessment

Chapter Test B

Atomic Physics

MULTIPLE CHOICE

In the space provided, write the letter of the term or phrase that best completes each statement or best answers each question.

_____ **1.** What is the frequency of a photon with an energy of 1.99×10^{-19} J? ($h = 6.63 \times 10^{-34}$ J•s)
  a. 1.00×10^{14} Hz   **c.** 3.00×10^{14} Hz
  b. 2.00×10^{14} Hz   **d.** 4.00×10^{14} Hz

_____ **2.** Light with an energy equal to three times the work function of a given metal causes the metal to eject photoelectrons. What is the ratio of the maximum photoelectron kinetic energy to the work function?
  a. $1 : 1$ **c.** $3 : 1$
  b. $2 : 1$ **d.** $4 : 1$

_____ **3.** A monochromatic light beam with a quantum energy value of 3.0 eV is incident upon a photocell. The work function of the photocell is 1.6 eV. What is the maximum kinetic energy of the ejected electrons?
  a. 4.6 eV   **c.** 1.4 eV
  b. 4.8 eV   **d.** 2.4 eV

_____ **4.** What causes the bright lines in the emission spectrum of an element to occur?
  a. Photons are absorbed when electrons jump from a higher-energy to a lower-energy state.
  b. Photons are emitted when electrons jump from a higher-energy to a lower-energy state.
  c. Photons are absorbed when electrons jump from a lower-energy to a higher-energy state.
  d. Photons are emitted when electrons jump from a lower-energy to a higher-energy state.

_____ **5.** What causes the dark lines in the absorption spectrum of an element to occur?
  a. Photons are absorbed when electrons jump from a higher-energy to a lower-energy state.
  b. Photons are emitted when electrons jump from a higher-energy to a lower-energy state.
  c. Photons are absorbed when electrons jump from a lower-energy to a higher-energy state.
  d. Photons are emitted when electrons jump from a lower-energy to a higher-energy state.

_____ **6.** What observation confirmed de Broglie's theory of matter waves?
  a. the photoelectric effect
  b. the scattering of alpha particles
  c. the diffraction of electrons
  d. the spontaneous emission of photons

_____ **7.** According to de Broglie, as the momentum of a moving particle is tripled, the corresponding wavelength changes by what factor?

  a. $\dfrac{1}{9}$ **c.** 3

  b. $\dfrac{1}{3}$ **d.** 9

_____ **8.** Why is a probability wave required to describe an electron's location?
  a. The electron's location can be precisely determined.
  b. Electrons violate Heisenberg's uncertainty principle.
  c. The electron may be found at various distances from the nucleus.
  d. The electron has less probability of being at the first Bohr orbit than at any other distance.

SHORT ANSWER

9. What was Planck's radical assumption about resonators in relation to black-body radiation?

_____

_____

_____

10. What are some weaknesses of Rutherford's atomic model?

_____

_____

_____

11. Starlight passes through a cloud of cool atomic gases. What kind of spectrum will be produced, and what will it look like?

_____

_____

_____

12. Which electron transitions in the Bohr hydrogen atom will produce photons with the shortest wavelengths?

_____

_____

_____

13. What causes an aurora, and why does it occur more easily and appear brighter nearer the poles than in equatorial or mid-latitude regions?

_____

_____

_____

14. You have designed an experiment to measure the momentum of light. Should you use light with long wavelengths or short wavelengths to better observe momentum transfer? Explain your answer.

_____

_____

_____

15. How does the classical model of standing waves on a vibrating string relate to the model of the Bohr atom in which electrons have specific de Broglie wavelengths?

_____

_____

_____

16. Why is the uncertainty principle more important for matter at the atomic level than for matter in large objects, such as a book or a car?

_____

_____

_____

| Chapter Test B continued

PROBLEM

17. What is the energy of a photon whose frequency is 5.0×10^{14} Hz?
$(h = 6.63 \times 10^{-34}$ J•s; 1 eV $= 1.60 \times 10^{-19}$ J)

18. What is the energy of a photon whose wavelength is 312 nm?
$(h = 6.63 \times 10^{-34}$ J•s; $c = 3.00 \times 10^8$ m/s; 1 eV $= 1.60 \times 10^{-19}$ J)

E_6 ——————————————— $E = -0.378$ eV
E_5 ——————————————— $E = -0.544$ eV
E_4 ——————————————— $E = -0.850$ eV

E_3 ——————————————— $E = -1.51$ eV

E_2 ——————————————— $E = -3.40$ eV

19. Calculate the frequency of the photon emitted when the electron in a hydrogen atom drops from energy level E_6 to energy level E_3 in the figure above. $(h = 6.63 \times 10^{-34}$ J•s; 1 eV $= 1.60 \times 10^{-19}$ J)

20. What is the de Broglie wavelength of a proton that has a mass of 1.67×10^{-27} kg and is moving at a speed of 2.7×10^5 m/s?
$(h = 6.63 \times 10^{-34}$ J•s)

Chapter Test A

Subatomic Physics

MULTIPLE CHOICE

In the space provided, write the letter of the term or phrase that best completes each statement or best answers each question.

_____ **1.** The nucleus of an atom is made up of which of the following combinations of particles?
    a. electrons and protons
    b. electrons and neutrons
    c. protons, electrons, and neutrons
    d. protons and neutrons

_____ **2.** To which of the following is the atomic number of a given element equivalent?
    a. the number of protons in the nucleus
    b. the number of neutrons in the nucleus
    c. the sum of the protons and neutrons in the nucleus
    d. the number of electrons in the outer shells

_____ **3.** Rutherford's experiments involving the use of alpha particle beams directed onto thin metal foils demonstrated the existence of which of the following?
    a. neutron   **c.** nucleus
    b. proton   **d.** positron

_____ **4.** What does the mass number of a nucleus indicate?
    a. the number of neutrons present
    b. the number of protons present
    c. the average atomic mass of the element
    d. the number of neutrons and protons present

_____ **5.** As the number of protons in the nucleus increases, the repulsive force
    a. becomes stronger.   **c.** remains unchanged.
    b. becomes weaker.   **d.** drops to zero.

_____ **6.** When are heavy nuclei most stable?
    a. when they contain more protons than neutrons
    b. when they contain more neutrons than protons
    c. when they contain equal numbers of protons and neutrons
    d. when the Coulomb force is stronger than the nuclear force

_____ **7.** How does a radioactive isotope that emits an alpha particle change?
  a. Atomic number decreases by four.
  b. Mass number decreases by four.
  c. Atomic number decreases by one.
  d. Mass number decreases by one.

_____ **8.** Of the main types of radiation emitted from naturally radioactive isotopes, which is the most penetrating?
  a. alpha   **c.** gamma
  b. beta   **d.** positron

_____ **9.** The alpha emission process results in the daughter nucleus differing in what manner from the parent?
  a. Atomic mass increases by one.
  b. Atomic number decreases by two.
  c. Atomic number increases by one.
  d. Atomic mass decreases by two.

_____ **10.** What particle is emitted when Pu-240 decays to U-236?
  a. alpha   **c.** positron
  b. beta   **d.** gamma

_____ **11.** Radium-226 decays to radon-222 by emitting
  a. beta particles.   **c.** gamma particles.
  b. alpha particles.   **d.** positrons.

_____ **12.** The natural logarithm of 2 (0.693) divided by the half-life of a radioactive substance is equal to the
  a. activity.   **c.** decay constant.
  b. decay rate.   **d.** decay lifetime.

_____ **13.** In fission reactions, how must the binding energy per nucleon vary?
  a. The binding energy per nucleon remains constant as atomic number increases.
  b. The binding energy per nucleon increases as atomic number increases.
  c. The binding energy per nucleon decreases as atomic number increases.
  d. none of the above

_____ **14.** In order to adequately control a chain reaction, it is necessary to have within the fissionable material a nonfissionable material. How does this material interact with neutrons?
  a. The material absorbs neutrons.
  b. The material emits neutrons.
  c. The material scatters neutrons.
  d. The material converts neutrons.

_____**15.** At this time, all nuclear reactors operate through
      a. fission only.
      b. fusion only.
      c. both fission and fusion.
      d. neither fission nor fusion.

_____**16.** How is a fission reactor different from a fusion reactor?
      a. The fuel is cheaper.
      b. The fuel must be processed.
      c. There is less radioactive waste.
      d. The transportation of fuel is safer.

_____**17.** Which interaction of nature is weakest?
      a. strong   **c.** electromagnetic
      b. weak   **d.** gravitational

_____**18.** Which interaction of nature binds neutrons and protons into nuclei?
      a. strong   **c.** electromagnetic
      b. weak   **d.** gravitational

_____**19.** Which of the following do physicists believe are fundamental particles?
      a. three quarks and three leptons
      b. six quarks and three leptons
      c. three quarks and six leptons
      d. six quarks and six leptons

_____**20.** Which statement about quarks is *not* correct?
      a. Only two quarks are needed to construct a hadron.
      b. An isolated quark has been observed by physicists.
      c. Every quark has an antiquark of opposite charge.
      d. There are six quarks that fit together in pairs.

SHORT ANSWER

21. What is half-life?

_____

_____

_____

22. What is nuclear fission?

_____

_____

23. According to the big bang theory, what occurred in the brief instant after the big bang?

_____

_____

_____

24. List the four fundamental interactions in order from weakest to strongest.

_____

_____

PROBLEM

25. Calculate the binding energy of the iron-56 nucleus.

$(c^2 = 931.49$ MeV/u; atomic mass of $^{56}_{26}$Fe $= 55.934\ 940$ u; atomic mass of $^{1}_{1}$H $= 1.007\ 825$ u; $m_n = 1.008\ 665$ u)

Chapter Test B

Subatomic Physics
MULTIPLE CHOICE

In the space provided, write the letter of the term or phrase that best completes each statement or best answers each question.

_____ 1. If there are 128 neutrons in Pb-210, how many neutrons are found in the nucleus of Pb-206?
     a. 122   **c.** 126
     b. 124   **d.** 130

_____ 2. What is the binding energy of a nucleus?
     a. the energy needed to remove one of the nucleons
     b. the average energy with which any nucleon is bound in the nucleus
     c. the energy released when nucleons bind together to form a stable nucleus
     d. the mass of the nucleus times c^2

_____ 3. If the stable nuclei are plotted with neutron number versus proton number, the curve formed by the stable nuclei does not follow the line $N = Z$. Which of the following influences the binding energy so that this "valley of stability" forms?
     a. the volume of the nucleus
     b. the size of the nuclear surface
     c. the Coulomb repulsive force
     d. the proton-neutron mass difference

_____ 4. In the radioactive formula, $^{220}_{86}Rn \rightarrow\ ^{216}_{84}Po + X$, what does X represent?
     a. $^{0}_{-1}e$ **c.** λ
     b. $^{0}_{1}e$ **d.** $^{4}_{2}He$

_____ 5. Samples of two different isotopes, X and Y, both contain the same number of radioactive atoms. The half-life of Sample X is twice that of Sample Y. How do their rates of radiation compare?
     a. Sample X has a greater rate than Sample Y.
     b. Sample X has a smaller rate than Sample Y.
     c. The rates of Sample X and Sample Y are equal.
     d. This cannot be determined from the information given.

_____ **6.** How many half-lives does it take for a radioactive substance to decay to 12.5 percent of its original amount?
   a. 1   **c.** 3
   b. 2   **d.** 4

_____ **7.** At around what mass number is the binding energy per nucleon greatest?
   a. 26   **c.** 111
   b. 58   **d.** 235

_____ **8.** Which of the following did *not* occur once the universe had cooled to a temperature of 3000 K?
   a. Quarks combined to form hadrons.
   b. Electrons and protons combined to form hydrogen atoms.
   c. Matter and radiation no longer interacted strongly.
   d. The removal of photon-scattering electrons made the universe "transparent."

_____ **9.** Which of the following is *not* one of the current questions regarding the standard model?
   a. Why do quarks carry fractional charge and electrons carry whole charge?
   b. What determines particle mass?
   c. How many quarks make up a proton?
   d. Can quarks exist in isolation?

SHORT ANSWER

10. List alpha, beta, and gamma radiations in order of decreasing speed.

_____

_____

11. Complete the following nuclear reaction.
   $^{230}_{90}\text{Th} \rightarrow {}^{226}_{88}\text{Ra} +$ _____.

_____

_____

_____

12. In the nuclear chain reaction of uranium-235, what particle reacts with the uranium nucleus to become a product of the reaction?

_____

13. In a fission reactor, what must be done to overcome the tendency of uranium-238 to absorb neutrons instead of undergoing fission?

_____

_____

_____

14. How is the neutron number for an isotope determined from the mass and atomic numbers?

_____

_____

_____

_____

15. How do leptons differ from hadrons?

_____

_____

_____

PROBLEM

16. Calculate the binding energy of the potassium-39 nucleus.

(c^2 = 931.49 MeV/u; atomic mass of $^{39}_{19}$K = 38.963 708 u; atomic mass of $^1_1$H = 1.007 825 u; m_n = 1.008 665 u)

17. Calculate the binding energy per nucleon of the gold-197 nucleus.

(c^2 = 931.49 MeV/u; atomic mass of $^{197}_{79}$Au = 196.966 543 u; atomic mass of $^1_1$H = 1.007 825 u; m_n = 1.008 665 u)

18. A radioactive material initially is observed to have an activity of 800.0 counts/s. If 4.0 h later it is observed to have an activity of 200.0 counts/s, what is its half-life?

19. Tritium (hydrogen-3) has a half-life of 12.3 years. How many years will have elapsed when the radioactivity of a tritium sample has decreased to 12.5 percent of its original value?

20. Three-fourths of a sample of iodine-131 decays in 16.14 days. What is the activity for a sample of iodine-131 that contains 5.5×10^{20} atoms?

Answer Key

The Science of Physics

CHAPTER TEST A (GENERAL)

1. b	8. c
2. c	9. b
3. b	10. c
4. b	11. d
5. a	12. a
6. c	13. d
7. d	14. c

15. d

$21.4 + 15 + 17.17 + 4.003 = \boxed{57.573}$

Answer rounds to 58 and is written as 5.8×10^1 in scientific notation.

16. b

17. c

18. a

19. c

20. b

21. Answers may vary. Sample answer: Mechanics studies the interactions of large objects, while quantum mechanics studies the behavior of subatomic (or very small) particles.

22. meter (m), kilogram (kg), and second (s)

23. *Solution*

$(92 \times 10^3 \text{ km})\left(\dfrac{10^4 \text{ dm}}{1 \text{ km}}\right) =$

$92 \times 10^7 \text{ dm} = \boxed{9.2 \times 10^8 \text{ dm}}$

24. the same dimensions (or units)

25. *Solution*   2 signif digits

$\dfrac{(8.86 + 1.0 \times 10^{-3})}{(3.610 \times 10^{-3})} =$?   sh be
                                                          2.5×10^3

$\dfrac{(8.86)}{(3.610 \times 10^{-3})} = \boxed{2.45 \times 10^3}$

The Science of Physics

CHAPTER TEST B (ADVANCED)

1. c

2. c

3. c

4. c

5. b

Solution

$6\,370\,000 \text{ m}\left(\dfrac{1 \text{ km}}{1000 \text{ m}}\right) =$

$\boxed{6.37 \times 10^3 \text{ km}}$

6. b

7. *Solution*

a

$(10.5) \times (8.8) \times (3.14) = \boxed{290.136}$

The answer rounds to 290 and is written as 2.9×10^2 in scientific notation.

8. a

$(0.82 + 0.042)(4.4 \times 10^3) =$

$(0.86)(4.4 \times 10^3) = \boxed{3784}$

The answer rounds to 3800 and is written as 3.8×10^3 in scientific notation.

9. d

10. b

11. *Solution*

d

$\dfrac{(\Delta v)^2}{\Delta x} = \dfrac{(\text{m/s})^2}{\text{m}} = \boxed{\text{m/s}^2}$

12. *Solution*

c

$\Delta x = Av$

Rearrange the equation to solve for A and substitute units.

$A = \dfrac{\Delta x}{v} = \dfrac{\text{m}}{\text{m/s}} = \boxed{\text{s}}$

13. b

Given

$m_{sun} = 2.0 \times 10^{30} \text{ kg}$

$m_{Hatom} = 1.67 \times 10^{-27} \text{ kg/atom}$

Solution

Estimate the answer using an order-of-magnitude calculation.

$10^{30}/10^{-27} = \boxed{10^{57}}$

14. Answers may vary. Sample answer: The model or hypothesis is unable to make reliable predictions.

15. The basic units can be combined to form derived units for other quantities.

16. 1×10^{-6} m

$1\mu\text{m}\dfrac{10^{-6} \text{ m}}{1\mu\text{m}} = \boxed{1 \times 10^{-6} \text{ m}}$

17. They return answers with as many digits as the display can show.

18. five

19. 10 000 or 10^4 Earths; 1.17×10^4 Earths

Given

$R_{Earth} = 6.37 \times 10^6$ m

Average Earth−sun distance =
$\quad 1.496 \times 10^{11}$

N_{Earths} between Earth and the sun $= ?$

Solution

$Diameter_{Earth} = 2(R_{Earth}) =$
$\quad 2(6.37 \times 10^6 \text{ m}) = 1.27 \times 10^7$ m

Therefore, using an order-of-magnitude calculation, the estimate for the number of Earths that would fit between Earth and the sun is

$$\frac{10^{11}}{10^7} = 10\ 000 \text{ or } 10^4.$$

Exact number of Earths is

$$\frac{(1.496 \times 10^{11} \text{ m})}{2(6.37 \times 10^6 \text{ m})} =$$

$\boxed{1.17 \times 10^4 \text{ Earths}}$

20. Yes; trial 2 has a much greater ΔT over the same period of time; temperature increases as time increases;
Since the table indicates a direct relationship between ΔT and Δt, the general form of the equation is $y = mx$. If ΔT is graphed on the y-axis and Δt is graphed on the x-axis, m represents the slope or $\dfrac{\Delta T}{\Delta t}$. In this instance, the average of $\dfrac{\Delta T}{\Delta t}$ is 0.12.
Therefore, the equation would be $\Delta T = 0.12\Delta t$.

Motion in One Dimension

CHAPTER TEST A (GENERAL)

1. a	**9.** d
2. d	**10.** a
3. c	**11.** c
4. b	**12.** c
5. c	**13.** c
6. d	**14.** a
7. c	**15.** a
8. a	**16.** c

17. displacement

18. The dog is moving at a constant speed because the position versus time graph is a straight line with a positive slope.

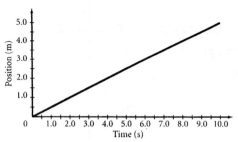

19. 3.3 m/s, to the right

Given

$x_i = -12$ m

$x_f = 24$ m

$\Delta t = 11$ s

Solution

$$v_{avg} = \frac{\Delta x}{\Delta t} = \frac{x_f - x_i}{\Delta t} =$$

$$\frac{(24 \text{ m}) - (-12 \text{ m})}{11 \text{ s}} =$$

$\boxed{3.3 \text{ m/s, to the right}}$

20. 44 m

Given

$a = -g = -9.81 \text{ m/s}^2$

$\Delta t = 2.0$ s

$v_i = -12$ m/s

Solution

$\Delta x = v_i\Delta t + \frac{1}{2}a(\Delta t)^2 = v_i\Delta t + \frac{1}{2}(-g)(\Delta t)^2$

$\Delta x = (-12.0 \text{ m/s})(2.0 \text{ s}) +$
$\quad \frac{1}{2}(-9.81 \text{ m/s}^2)(2.0 \text{ s})^2 = \boxed{-44\text{m}}$

Motion in One Dimension

CHAPTER TEST B (ADVANCED)

1. a	**6.** b
2. b	**7.** a
3. c	**8.** c
4. a	**9.** d
5. b	**10.** c

11. b

12. Although the magnitudes of the displacements are equal, the displacements are in opposite directions. Therefore, one displacement is positive and one displacement is negative.

13. The displacement is negative because a change of position in the direction opposite of increasing positive position is negative displacement.

14. The dog's initial position and its final position are the same position.

15. Since the usual choice of coordinates uses positive as the direction away from Earth, the direction of free-fall acceleration is negative because the object accelerates toward Earth.

16. 1.7×10^{-2} h

Given

$v_{avg} = 1.8$ km/h

$\Delta x = 0.30$ km

Solution

$v_{avg} = \dfrac{\Delta x}{\Delta t}$

$\Delta t = \dfrac{\Delta x}{v_{avg}} = \dfrac{0.30 \text{ km}}{18 \text{ km/h}} = \boxed{1.7 \times 10^{-2} \text{ h}}$

17. 1.2 km, north

Given

$v_{avg,1} = -0.75$ km/h

$\Delta t_1 = 1.5$ h

$v_{avg,2} = 0.90$ km/h

$\Delta t_2 = 2.5$ h

Solution

$\Delta x = \Delta x_1 + \Delta x_2 = v_{avg,1}\Delta t_1 + v_{avg,2}\Delta t_2$

$\Delta x = (-0.75 \text{ km/h})(1.5 \text{ h}) + (0.90 \text{ km/h})(2.5 \text{ h}) = \boxed{1.2 \text{ km, north}}$

18. 1.0 m/s

Given

$v_i = 1.8$ m/s

$a = -3.00$ m/s²

$\Delta x = 0.37$ m

Solution

$v_f^2 = v_i^2 + 2a\Delta x$

$v_f = \sqrt{v_i^2 + 2a\Delta x} =$

$\sqrt{(1.8 \text{ m/s})^2 + (2)(-3.0 \text{ m/s}^2)(0.37 \text{ m})}$

$\boxed{v_f - 1.0 \text{ m/s}}$

19. at least 0.20 m

Given

$a = -g = -9.81$ m/s²

$\Delta t = 0.20$ s

$v_i = 0.0$ m/s

Solution

$\Delta x = v_i\Delta t + \tfrac{1}{2}a(\Delta t)^2 =$

$v_i\Delta t + \tfrac{1}{2}(-g)(\Delta t)^2$

$\Delta x = (0 \text{ m/s})(0.20 \text{ s}) +$

$\tfrac{1}{2}(-9.81 \text{ m/s}^2)(0.20 \text{ s})^2 = \boxed{-0.20 \text{ m}}$

20. 30.5 m

Given

$a = -g = -9.81$ m/s²

$v_{i,1} = 0.0$ m/s

$\Delta x = -32.0$ m

$v_{i,1} = 0.0$ m/s

$\Delta t_{1,2} = 2.0$ s

Solution

$\Delta x_1 = v_{i,1}\Delta t_1 + \tfrac{1}{2}a(\Delta t_1)^2 =$

$v_{i,1}\Delta t_1 + \tfrac{1}{2}(a)(\Delta t_1)^2$

$\Delta t_1 = \sqrt{\dfrac{2\Delta x_1}{a}} = \sqrt{\dfrac{2\Delta x_1}{-g}} =$

$\sqrt{\dfrac{2(-32.0 \text{ m})}{-9.81 \text{ m/s}^2}} = 2.56$ s

$\Delta t_2 = \Delta t_1 - \Delta t_{1,2} = 2.56 \text{ s} - 2.00 \text{ s} = 0.56$ s

$\Delta x_2 = v_{i,2}\Delta t_2 + \tfrac{1}{2}a(\Delta t_2)^2 = v_{i,2}\Delta t_2 + \tfrac{1}{2}(-g)(\Delta t_2)^2$

$\Delta x_2 = (0.0 \text{ m/s})(0.56 \text{ s}) +$

$\tfrac{1}{2}(-9.81 \text{ m/s}^2)(0.56 \text{ s})^2 = -1.5$ m

$h = 32.0 \text{ m} - 1.5 \text{ m} = \boxed{30.5 \text{ m}}$

Two-Dimensional Motion and Vectors

CHAPTER TEST A (GENERAL)

1. b
2. a
3. b
4. d
5. a
6. a
7. c
8. b
9. d
10. b
11. b
12. a
13. c
14. b
15. c
16. a

17. Displacement is a vector quantity.

18. The vectors must be perpendicular to each other.

19. 120 m

Given

$v_i = 12$ m/s at 30.0° above the horizontal

$\Delta t = 5.6$ s

$q = 9.81$ m/s²

Solution

$$v_{i,y} = v_i \sin \theta = (12 \text{ m/s})(\sin 30.0°) = 6.0 \text{ m/s}$$

$$\Delta y = \frac{1}{2}a_y(\Delta t)^2 + v_{i,y}\Delta t =$$

$$\frac{1}{2}(-g)(\Delta t)^2 + v_{i,y}\Delta t =$$

$$\frac{1}{2}(-9.81 \text{ m/s}^2)(5.6 \text{ s})^2 + (6.0 \text{ m/s})(5.6 \text{ s})$$

$$\Delta y = -120 \text{ m}$$

$$h = \boxed{120 \text{ m}}$$

20. 27 km/h north

Given

v_{pg} = velocity of plane to ground = 145 km/h south

v_{pa} = velocity of plane to air = 170.0 km/h south

Solution

$$v_{pg} = v_{pa} + v_{ag}$$
$$v_{ag} = v_{pg} - v_{pa}$$
$$v_{ag} = 145 \text{ km/h} - 172 \text{ km/h} = -27 \text{ km/h}$$

$$v_{ag} = \boxed{27 \text{ km/h north}}$$

Two-Dimensional Motion and Vectors

CHAPTER TEST B (Advanced)

1. b

2. d

3. d

Given

$\Delta x_1 = 3.0 \times 10^1$ cm east

$\Delta y_1 = 25$ cm north

$\Delta x_2 = 15$ cm west

Solution

$$\Delta x_{tot} = \Delta x_1 + \Delta x_2 = (3.0 \times 10^1 \text{ cm}) + (-15 \text{ cm}) = 15 \text{ cm}$$

$$\Delta y_{tot} = \Delta y_2 = 25 \text{ cm}$$

$$d^2 = (\Delta x_{tot})^2 + (\Delta y_{tot})^2$$

$$d = \sqrt{(\Delta x_{tot})^2 + (\Delta y_{tot})^2} = \sqrt{(15 \text{ cm})^2 + (25 \text{ cm})^2}$$

$$d = \boxed{29 \text{ cm}}$$

4. a

5. d

Solution

$$\Delta x_1 = 2.00 \times 10^2 \text{ units}$$

$$\Delta y_1 = 0$$

$$\Delta x_2 = d_2 \cos \theta = (4.00 \times 10^2 \text{ units})(\cos 30.0°) = 3.46 \times 10^2 \text{ units}$$

$$\Delta y_2 = d_2 \sin \theta = (4.00 \times 10^2 \text{ units})(\sin 30.0°) = 2.00 \times 10^2 \text{ units}$$

$$\Delta x_{tot} = \Delta x_1 + \Delta x_2 = (2.00 \times 10^2 \text{ units}) - (3.46 \times 10^2 \text{ units}) = -1.46 \times 10^2 \text{ units}$$

$$\Delta y_{tot} = \Delta y_1 + \Delta y_2 = 0 + (2.00 \times 10^2 \text{ units}) = 2.00 \times 10^2 \text{ units}$$

$$d^2 = (\Delta x_{tot})^2 + (\Delta y_{tot})^2$$

$$d = \sqrt{(\Delta x_{tot})^2 + (\Delta y_{tot})^2} = \sqrt{(-1.46 \times 10^2 \text{ units})^2 + (2.00 \times 10^2 \text{ units})^2}$$

$$d = 2.48 \times 10^2 \text{ units}$$

$$\theta = \tan^{-1}\left(\frac{\Delta y}{\Delta y}\right) =$$

$$\tan^{-1}\frac{2.00 \times 10^2 \text{ units}}{1.46 \times 10^2 \text{ units}} = 53.9°$$

$$d = \boxed{2.48 \times 10^2 \text{ units } 53.9° \text{ north of west}}$$

6. b

7. d

8. b

Solution

$$v_x = v_{i,x} = v_i \cos \theta = (12 \text{ m/s})(\cos 20.0°) = 11 \text{ m/s}$$

$$v_{i,y} = v_i \sin \theta = (12 \text{ m/s})(\sin 20.0°) = 4.1 \text{ m/s}$$

$$\Delta y = 0 = \frac{1}{2}a_y(\Delta t)^2 + v_{i,y}\Delta t$$

$$\Delta t = \frac{-2v_{i,y}}{a_y} = \frac{-2v_{i,y}}{-g} = \frac{-2(4.1 \text{ m/s})}{-9.81 \text{ m/s}}$$

$$= 0.84 \text{ s}$$

$$\Delta x = v_x\Delta t = (11 \text{ m/s})(0.84 \text{ s}) = \boxed{9.2 \text{ m}}$$

9. c

10. b

Given

v_{pa} = velocity of plane relative to the air = 500.0 km/h east

v_{ag} = velocity of air relative to the ground = 120.0 km/h 30.00° north of east

Solution

$$v_{ag,x} = v_{ag} \cos \theta = (120.0 \text{ km/h})$$
$$(\cos 30.00°) = 103.9 \text{ km/h}$$

$$v_{ag,y} = v_{ag} \sin \theta = (120.0 \text{ km/h})$$
$$(\sin 30.00°) = 60.0 \text{ km/h}$$

$$v_{pg,x} = v_{pa} + v_{ag,x} = 500.0 \text{ km/h} +$$
$$103.9 \text{ km/h} = 603.9 \text{ km/h}$$

$$v_{pg,y} = 60.0 \text{ km/h}$$

$$v_{pg} = \sqrt{(v_{pg,x})^2 + (v_{pg,y})^2} =$$

$$\sqrt{(603.9 \text{ km/h})^2 + (60.0 \text{ km/h})^2} =$$

$$\boxed{606.9 \text{ km/h}}$$

11. The triangle method of adding vectors requires that you align the vectors, one after the other, tail to nose, by moving them parallel and perpendicular to their original orientations. The resultant vector is an arrow drawn from the tail of the first vector to the tip of the last vector.

12. The magnitude of the other component vector is zero.

13. Resolve each vector into perpendicular components and add the components that lie along the same axis. The resultant vectors can be added by using the Pythagorean theorem because they are perpendicular.

14. Objects sent into the air and subject to gravity exhibit projectile motion.

15. 12.2 m

Solution

$$d = \sqrt{(12.0 \text{ m})^2 + (2.5 \text{ m})^2} = \boxed{12.2 \text{ m}}$$

16. 62 steps

Solution

$$d = \sqrt{(28 \text{ steps})^2 + (55 \text{ steps})^2} =$$

$$\boxed{62 \text{ steps}}$$

17. 43 m

Given

$$\Delta x_1 = -1.0 \times 10^1 \text{ m}$$
$$\Delta y_1 = 15 \text{ m}$$
$$\Delta x_2 = +5.0 \times 10^1 \text{ m}$$

Solution

$$\Delta x_{tot} = \Delta x_1 + \Delta x_2 = (-1.0 \times 10^1 \text{ m})$$
$$+ (5.0 \times 10^1 \text{ m}) = 4.0 \times 10^1 \text{ m}$$

$$\Delta y_{tot} = \Delta y_1 = -15 \text{ m}$$

$$d^2 = (\Delta x_{tot})^2 + (\Delta y_{tot})^2$$

$$d = \sqrt{(\Delta x_{tot})^2 + (\Delta y_{tot})^2} =$$

$$\sqrt{(4.0 \times 10^1 \text{ m})^2 + (1.5 \times 10^1 \text{ m})^2}$$

$$= \boxed{4.3 \times 10^1 \text{ m}}$$

18. 4.9 m

Given

d_1 = 3.2 m along +y-axis

d_2 = 4.6 m at 195° counterclockwise from +x-axis

$$d_1 = 3.2 \text{ m} \qquad \theta_1 = 0.0°$$
$$d_2 = 4.6 \text{ m} \qquad \theta_2 = 195°$$

Solution

$$\Delta x_1 = 0.0 \text{ m}$$
$$\Delta y_1 = 3.2 \text{ m}$$
$$\Delta x_2 = d_2 \cos \theta = (4.6 \text{ m})(\cos 195°) = -4.4 \text{ m}$$
$$\Delta y_2 = d_2 \sin \theta = (4.6 \text{ m})(\sin 195°) = -1.2 \text{ m}$$
$$\Delta x_{tot} = \Delta x_1 + \Delta x_2 = (0 \text{ m}) + (-4.4 \text{ m}) = -4.4 \text{ m}$$
$$\Delta y_{tot} = \Delta y_1 + \Delta y_2 = (3.2 \text{ m}) + (-1.2 \text{ m}) = 2.0 \text{ m}$$
$$d^2 = (\Delta x_{tot})^2 + (\Delta y_{tot})^2$$
$$d = \sqrt{(\Delta x_{tot})^2 + (\Delta y_{tot})^2} =$$
$$\sqrt{(-4.4 \text{ m})^2 + (2.0 \text{ m})^2} = \boxed{4.9 \text{ m}}$$

19. 226 m

Given

v = 50.0 m/s horizontally

Δy = −100.0 m

Solution

$\mathbf{v_{i,x}} = \mathbf{v_x}$ = 50.0 m/s horizontally

$\mathbf{v_{i,y}} = 0$

$$\Delta y = \frac{1}{2} a_y (\Delta t)^2$$

$$(\Delta t)^2 = \frac{2 \Delta y}{a_y}$$

$$\Delta t = \sqrt{\frac{2 \Delta y}{a_y}} = \sqrt{\frac{2 \Delta y}{(-g)}} =$$

$$\sqrt{\frac{2(-100.0 \text{ m})}{-9.81 \text{ m/s}^2}} = 4.52 \text{ s}$$

$$\Delta x = v_x \Delta t = (50.0 \text{ m/s})(4.52 \text{ s}) =$$

$$\boxed{226 \text{ m}}$$

20. 208 s

Given

$\mathbf{v_{rg}}$ = velocity of river to ground = 2.00 m/s downstream

$\mathbf{v_{br}}$ = velocity of boat to river = 10.00 m/s

x_1 = 1000.0 m downstream

x_2 = −1000.0 m downstream

$\mathbf{v_{bg}}$ = velocity of boat

Solution

downstream

$\mathbf{v_{bg}} = \mathbf{v_{br}} + \mathbf{v_{rg}}$

$v_{bg} = 10.00 \text{ m/s} + 2.00 \text{ m/s} = 12.00 \text{ m/s}$

$\Delta t_1 = \Delta x_1 / v_{bg} = 1000.0 \text{ m} / 12.00 \text{ m/s} = 83.33 \text{ s}$

upstream

$\mathbf{v_{bg}} = \mathbf{v_{br}} + \mathbf{v_{rg}}$

$v_{bg} = -10.00 \text{ m/s} + 2.00 \text{ m/s} = -8.00 \text{ m/s}$

$\Delta t_2 = \Delta x_2 \backslash v_{bg} = \dfrac{-1000.0 \text{ m}}{-8.00 \text{ m/s}} = 125 \text{ s}$

$\Delta t = \Delta t_1 + \Delta t_2 = 83.33 \text{ s} + 125 \text{ s} = \boxed{208 \text{ s}}$

Forces and the Laws of Motion

CHAPTER TEST A (General)

1. c	10. d
2. d	11. c
3. d	12. a
4. c	13. d
5. c	14. d
6. c	15. b
7. c	16. d
8. b	17. c
9. d	18. d

19. Forces exerted by the object do not change its motion.

20. An object at rest remains at rest and an object in motion continues in motion with constant velocity unless it experiences a net external force.

21. $\Sigma \mathbf{F}$ is the vector sum of the external forces acting on the object.

22. In most cases, air resistance increases with increasing speed.

23. 27 N, to the right

Given

$\mathbf{F_1} = 102 \text{ N, to the right}$

$\mathbf{F_2} = 75 \text{ N, to the left}$

Solution

$\mathbf{F_{net}} = \mathbf{F_1} + \mathbf{F_2}$

$F_{net} = F_1 - F_2 = 102 \text{ N} - 75 \text{ N} = 27 \text{ N}$

$\mathbf{F_{net}} = \boxed{27 \text{ N, to the right}}$

24. 16 N

Given

$m = 33 \text{ kg}$

$a = 0.50 \text{ m/s}^2$

Solution

$F_{net, x} = \Sigma F_x = ma_x = (32 \text{ kg})(0.50 \text{ m/s}^2) = \boxed{16 \text{ N}}$

25. 10.4 N

Given

$m = 1.10 \text{ kg}$

$\alpha = 15.0°$

$g = 9.81 \text{ m/s}^2$

Solution

$F_{net, y} = \Sigma F_y = F_n - F_y = 0$

$\theta = 180° - 90° - 15.0° = 75.0°$

$F_n = F_y = F_g \sin \theta = mg \sin \theta$

$F_n = (1.10 \text{ kg})(9.81 \text{ m/s}^2)(\sin 75.0°)$

$= \boxed{10.4 \text{ N}}$

Forces and the Laws of Motion

CHAPTER TEST B (Advanced)

1. d

2. a

3. c

4. b

Given

$F_y = 60.0 \text{ N}$

$\theta = 30.0°$

Solution

$\cos \theta = \dfrac{F_y}{F}$

$F = \dfrac{F_y}{\cos \theta} = \dfrac{60.6 \text{ N}}{\cos 30.0°} = \boxed{70.0 \text{ N}}$

5. c	8. a
6. d	9. c
7. d	10. a

11. b

12. a

Given

$F_g = 1.0 \times 10^2 \text{ N}$

$\theta = 20.0°$

Solution

$\Sigma F_y = F_n - F_{g,y} = 0$

$F_n = F_{g, y} = F_g \cos \theta = (1.0 \times 10^2 \text{ N})(\cos 20.0°) = \boxed{94 \text{ N}}$

13. b

Given

$F_{g, book} = 5 \text{ N}$

$\mu_s = 0.2$

Solution

$$F_{net} = \Sigma F_x = F_{applied} - F_{s,\,max} = 0$$
$$F_{applied} = F_{s,\,max} = \mu_s F_n = \mu_s F_g$$
$$F_g = (5\text{ N} + 5\text{ N} + 5\text{ N} + 5\text{ N} + 5\text{ N})$$
$$= 25\text{ N}$$

$$F_{applied} = (0.2)(25\text{ N}) = \boxed{5\text{ N}}$$

14. Force causes an acceleration or a change in an object's velocity. Applying the brakes decelerates the bicycle (accelerates it in the negative direction) and causes a change in the bicycle's velocity because the bicycle slows down.

15. A scalar quantity has only magnitude. Force has both magnitude and direction; so, it cannot be a scalar quantity.

16.

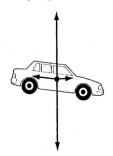

17. The natural condition for a moving object is to remain in motion once it has been set in motion.

18. Gravity exerts a downward force on the car that is balanced by the normal force of the road acting upward on the car. The car's forward motion is opposed by the friction between the road and the tires and by the resistance of the air. The sum of these opposing forces is balanced by an equal and opposite force exerted by the engine and applied to the tires, where the road exerts a reaction force that is directed forward.

19. Mass is the amount of matter in an object and is an inherent property of an object. Weight is not an inherent property of an object and is the magnitude of the force due to gravity acting on the object.

20. The acceleration is then zero, and the car moves at a constant speed.

21. Air resistance is a form of friction because it is a retarding force. It acts in the direction opposite an object's motion.

22. 48 N

Given
$$F = 8.0 \times 10^1\text{ N}$$
$$\theta = 53°$$

Solution
$$\Sigma F = F_f - F_y = 0$$
$$F_f = F_y = F \cos\theta = (8.0 \times 10^1\text{ N})$$
$$(\cos 53°) = \boxed{48\text{ N}}$$

23. 1.3 m/s², upward

Given
$$F_{applied,\,y} = 277\text{ N}$$
$$F_g = 245\text{ N}$$
$$m = 25\text{ kg}$$

Solution
$$F_{net} = \Sigma F_y = F_{applied,\,y} - F_g = ma$$
$$a = \frac{(F_{applied,\,y} - F_g)}{m} =$$
$$\frac{(277\text{ N} - 245\text{ N})}{(25\text{ kg})} = 1.3\text{ m/s}^2$$
$$\mathbf{a} = \boxed{1.3\text{ m/s}^2,\text{ upward}}$$

24. 14 N, upward

Given
$$F_{g,\,1} = 5\text{ N}$$
$$F_{g,\,2} = 9\text{ N}$$
$$F_{g,\,3} = 16\text{ N}$$

Solution
$$F_{net,\,y} = \Sigma F_y = F_n - F_{g,\,1} - F_{g,\,2} = 0$$
$$F_n = F_{g,\,1} + F_{g,\,2} = 5\text{ N} + 9\text{ N} = 14\text{ N}$$
$$\mathbf{F_n} = \boxed{14\text{ N, upward}}$$

25. 0.387

Given
$$m = 1.00 \times 10^2\text{ kg}$$
$$a_x = 0.70\text{ m/s}^2$$
$$\theta = 25.0°$$
$$g = 9.81\text{ m/s}^2$$

Solution
$$\Sigma F_x = F_x - F_f = F_{net} = ma_x$$
$$F_f = F_x - F_{net}$$
$$F_{net} = ma_x$$
$$\mu_k F_n = \mu_k\, mg \cos\theta =$$
$$\mu_k\, (1.00 \times 10^2\text{ kg})(9.81\text{ m/s}^2)(\cos 25.0°)$$
$$\mu_k F_n = \mu_k\, (8.89 \times 10^2\text{ N})$$
$$F_x = mg \sin\theta = (1.00 \times 10^2\text{ kg})$$
$$(9.81\text{ m/s}^2)(\sin 25°) = 4.14 \times 10^2\text{ N}$$
$$F_{net} = ma_x = (1.00 \times 10^2\text{ kg})$$
$$(0.70\text{ m/s}^2) = 0.70 \times 10^2\text{ N}$$

$$\mu_k = \frac{F_x}{F_n} - \frac{F_{net}}{F_n} =$$

$$\frac{(4.14 \times 10^2 \text{ N} - 0.70 \times 10^2 \text{ N})}{(8.89 \times 10^2 \text{ N})} =$$

$$\boxed{0.387}$$

Work and Energy

CHAPTER TEST A (GENERAL)

1. c	11. d
2. c	12. b
3. b	13. d
4. a	14. d
5. b	15. d
6. c	16. d
7. b	17. c
8. d	18. c
9. b	19. d
10. c	20. b

21. The net work is zero (because the net force on the car is zero).
22. The net work done by the net force acting on an object is equal to the change in the kinetic energy of the object.
23. 5.0 J
24. 215 J

Given
$F = 43.0$ N
$d = 5.00$ m

Solution
$W = Fd = (43.0 \text{ N})(5.00 \text{ M}) = \boxed{215 \text{ J}}$

25. 2.5 kW

Given
$m = 250$ kg
$d = 2.0$ m
$\Delta t = 2.0$ s
$g = 9.81$ m/s^2

Solution
$$P = \frac{W}{\Delta t} = \frac{Fd}{\Delta t} = \frac{mgd}{\Delta t} =$$
$$\frac{(250 \text{ kg})(9.81 \text{ m/s}^2)(2.0 \text{ m})}{2.0 \text{ s}} =$$

$$2.5 \times 10^3 \text{ W} = \boxed{2.5 \text{ kW}}$$

Work and Energy

CHAPTER TEST B (ADVANCED)

1. b
2. a
3. b
4. d

5. c

Solution
$$KE = \frac{1}{2}mv^2 = \frac{1}{2}(0.1335 \text{ kg})(40.0 \text{ m/s}^2)$$
$$= \boxed{108 \text{ J}}$$

6. a
7. d
8. b
9. b

Solution
$PE = mgh =$
$$(1.0 \text{ kg})(9.81 \text{ m/s}^2)(1.0 \text{ m}) = \boxed{9.8 \text{ J}}$$

10. a
11. d
12. c

Solution
$KE_f = PE_{g,i} = mgh =$
$(3.00 \text{ kg})(9.81 \text{ m/s}^2)(1.00 \text{ m}) = \boxed{29.4 \text{ J}}$

13. c

Solution
$$P = \frac{W}{\Delta t} = \frac{Fd}{\Delta t} = \frac{mgd}{\Delta t} =$$
$$\frac{(60.0 \text{ kg})(9.81 \text{ m/s}^2)(4.0 \text{ m})}{4.2 \text{ s}} =$$

$$\boxed{5.6 \times 10^2 \text{ W}}$$

14. Work, in the scientific sense, is the product of the component of a force along the direction of displacement and the magnitude of the displacement. No work is done unless a force causes some displacement that is not perpendicular to the force.
15. gravitational potential energy
16. At the top of the fall, all the energy is gravitational potential energy. During the fall, gravitational potential energy decreases as it is transformed into kinetic energy. When the pencil reaches the ground, all the energy is kinetic energy.
17. $\frac{1}{2}mv_i^2 + mgh_i + \frac{1}{2}kx_i^2 = \frac{1}{2}mv_f^2 + mgh_f + \frac{1}{2}kx_f^2$
18. Definition of *power*: $P = \dfrac{W}{\Delta t}$
Definition of *work*: $W = Fd$

Definition of *speed*: $v = \dfrac{d}{\Delta t}$

$P = \dfrac{W}{\Delta t} = \dfrac{Fd}{\Delta t} = Fv$
Alternative definition of power: $P = Fv$

19. The 20 kW motor does twice as much work in the same amount of time.

20. 4.7×10^5 J

Given

$m = 1.5 \times 10^5$ J

$v = 25$ m/s

Solution

$KE = \frac{1}{2}mv^2 = \frac{1}{2}(1.5 \times 10^5 \text{ J})(25 \text{ m/s})^2$

$= \boxed{4.7 \times 10^5 \text{ J}}$

21. 35 J

Given

$F_w = 50.0$ N

$F_k = -43$ N

$d = 5.0$ m

Solution

$W_{net} = F_{net}d = (F_w + F_k)d = [(50.0 \text{ N})$

$+ (-43 \text{ N})]\,(5.0 \text{ m}) = \boxed{35 \text{ J}}$

22. 320 m

Given

$v_i = 0$ m/s

$v_f = 56$ m/s

$\theta = 30.0°$

$g = 9.81$ m/s^2

Solution

$W_{net} = \Delta KE$

$W_{net} = Fd = (F_g \sin \theta)d = mgd \sin \theta$

$\Delta KE = KE_f - KE_i = \frac{1}{2}mv^2_f - 0 =$

$\frac{1}{2}mv^2_f$

$mgd \sin \theta = \frac{1}{2}mv^2_f$

$d = \dfrac{v^2_f}{2g \sin \theta}$

$d = \dfrac{(56 \text{ m/s})^2}{2(9.81 \text{ m/s}^2)(\sin 30.0°)} =$

$\boxed{3.2 \times 10^2 \text{ m}}$

23. 6.94×10^6 J

Given

$m = 80.0$ kg

$h = 8848$ m

$g = 9.81$ m/s^2

Solution

$PE = mgh = (80.0 \text{ kg})(9.81 \text{ m/s}^2)$

$(8848 \text{ m}) = \boxed{6.94 \times 10^6 \text{ J}}$

24. 28.0 m/s

Given

$h = 40.0$ m

$g = 9.81$ m/s^2

Solution

$KE_i = PE_{g,f}$

$\frac{1}{2}mv_i^2 = mgh$

$v_i = \sqrt{2gh} =$

$\sqrt{(2)(9.81 \text{ m/s}^2)(40.0 \text{ m})} = \boxed{28.0 \text{ m/s}}$

25. 589 MW

Given

$flow = 1.20 \times 10^6 \text{ kg/s} = \dfrac{m}{\Delta t}$

$d = 50.0$ m

$g = 9.81$ m/s^2

Solution

$\dfrac{m}{\Delta t} = 1.20 \times 10^6$ kg/s

$P = \dfrac{W}{\Delta t} = \dfrac{Fd}{\Delta t} = \dfrac{mgd}{\Delta t} = \dfrac{m}{\Delta t} gd$

$P = (1.20 \times 10^6 \text{ kg/s})(9.81 \text{ m/s}^2)$

$(50.0 \text{ m}) = 5.89 \times 10^8 \text{ W} = \boxed{589 \text{ MW}}$

Momentum and Collisions

CHAPTER TEST A (GENERAL)

1. c

2. c

3. b

4. c

5. a

Given

$\mathbf{p_i} = 4.0$ kg•m/s

$\mathbf{p_f} = -4.0$ kg•m/s

Solution

$\Delta\mathbf{p} = \mathbf{p_f} - \mathbf{p_i} = (-4.0 \text{ kg•m/s}) -$

$4.0 \text{ kg•m/s} = -8.0 \text{ kg•m/s}$

6. c

7. b

8. b

9. a

10. d

11. d

12. a

13. d

14. c

15. The bullet's momentum decreases as its speed decreases.

16. The student has the least momentum when dodging the opening door.

17. A small force can produce a large change in momentum if the force acts on an object for a long period of time.

18. zero

19. They have the same momentum.
$(1.85 \times 10^4 \text{ kg} \cdot \text{m/s})$
Given
$m_1 = 6160 \text{ kg}$
$\mathbf{v_1} = 3.00 \text{ m/s}$
$m_2 = 1540 \text{ kg}$
$\mathbf{v_2} = 12.0 \text{ m/s}$
Solution
$\mathbf{p_1} = m_1\mathbf{v_1} = (6160 \text{ kg})(3.00 \text{ m/s}) =$
$1.85 \times 10^4 \text{ kg} \cdot \text{m/s}$
$\mathbf{p_2} = m_2\mathbf{v_2} = (1540 \text{ kg})(12.0 \text{ m/s}) =$
$1.85 \times 10^4 \text{ kg} \cdot \text{m/s}$

$\boxed{\mathbf{p_1} = \mathbf{p_2}}$

20. $-1.2 \text{ kg} \cdot \text{m/s}$
Given
$m = 0.15 \text{ kg}$
$\mathbf{v_i} = 5.0 \text{ m/s}$
$\mathbf{v_f} = -3.0 \text{ m/s}$

Solution
$\Delta\mathbf{p} = m(\mathbf{v_f} - \mathbf{v_i}) =$
$(0.15 \text{ kg})(-3.0 \text{ m/s} - 5.0 \text{ m/s}) =$
$\boxed{-1.2 \text{ kg} \cdot \text{m/s}}$

Momentum and Collisions

CHAPTER TEST B (ADVANCED)

1. a
Given
a: $m = 275 \text{ kg}$
$\mathbf{v} = 0.55 \text{ m/s}$
b: $m = 2.7 \text{ kg}$
$\mathbf{v} = 7.5 \text{ m/s}$
c: $m = 91 \text{ kg}$
$\mathbf{v} = 1.4 \text{ m/s}$
d: $m = 1.8 \text{ kg}$
$\mathbf{v} = 6.7 \text{ m/s}$

Solution
$\mathbf{p} = m\mathbf{v}$
$\mathbf{p_a} = (275 \text{ kg})(0.55 \text{ m/s}) =$
$1.5 \times 10^2 \text{ kg} \cdot \text{m/s}$
$\mathbf{p_b} = (2.7 \text{ kg})(7.5 \text{ m/s}) =$
$2.0 \times 10^1 \text{ kg} \cdot \text{m/s}$
$\mathbf{p_c} = (91 \text{ kg})(1.4 \text{ m/s}) =$
$1.3 \times 10^2 \text{ kg} \cdot \text{m/s}$
$\mathbf{p_d} = (1.8 \text{ kg})(6.7 \text{ m/s}) =$
$1.2 \times 10^1 \text{ kg} \cdot \text{m/s}$

$\boxed{\mathbf{p_a} > \mathbf{p_c} > \mathbf{p_b} > \mathbf{p_d}}$

2. a

3. d
Given
$m = 2.0 \text{ kg}$
$\mathbf{v_i} = 40 \text{ m/s}$
$\mathbf{v_f} = -60 \text{ m/s}$
Solution
$\Delta\mathbf{p} = m(\mathbf{v_f} - \mathbf{v_i}) =$
$(0.2 \text{ kg})(-60 \text{ m/s} - 40 \text{ m/s}) =$
$\boxed{-20 \text{ kg} \cdot \text{m/s}}$

4. b
5. a
6. b
7. b
8. d
9. b
10. The first pitch is harder to stop. The first pitch has greater momentum because it has a greater velocity.
11. Yes, a spaceship traveling with constant velocity could experience a change in momentum if its mass changed, for example, by burning fuel.
12. 10.5 m/s
13. Stopping a falling egg requires changing the momentum of the egg from its value at the time of first impact to zero. If the egg hits the concrete, the time interval over which this happens is very small, so the force is large. If the egg lands on grass, the time interval over which the momentum changes is larger, so the force on the egg is smaller.
14. Producing sound requires energy. Because the system of objects loses some energy as sound is produced in the collision, the total kinetic energy cannot be conserved, so the collision cannot be perfectly elastic.
15. 30 m/s to the west
Given
$m_1 = 2680 \text{ kg}$
$\mathbf{v_1} = 15 \text{ m/s to the west}$
$m_2 = 1340 \text{ kg}$
Solution
$m_1\mathbf{v_1} = m_2\mathbf{v_2}$
$\mathbf{v_2} = \dfrac{m_1\mathbf{v_1}}{m_2} = \dfrac{(2.68 \times 10^3 \text{ kg})(15 \text{ m/s})}{(1.34 \times 10^3 \text{ kg})}$
$= \boxed{3.0 \times 10^1 \text{ m/s}}$

16. -1.8 kg•m/s

Given

$m = 6.0 \times 10^{-2}$ kg

$\mathbf{v_i} = 12$ m/s

$\mathbf{v_f} = -18$ m/s

Solution

$\Delta p = m(\mathbf{v_f} - \mathbf{v_i}) =$

$(6.0 \times 10^{-2}$ kg$)(-18$ m/s $- 12$ m/s$)$

$= \boxed{-1.8 \text{ kg•m/s}}$

17. 77 s; 5.8×10^2 m

Given

$m = 1.8 \times 10^5$ kg

$\mathbf{v_i} = 15$ m/s

$\mathbf{v_f} = 0$ m/s

$\mathbf{F} = -3.5 \times 10^4$ N

Solution

$\mathbf{F}\Delta t = \Delta \mathbf{p}$

$\Delta t = \dfrac{\Delta \mathbf{p}}{\mathbf{F}} = \dfrac{m(\mathbf{v_f} - \mathbf{v_i})}{\mathbf{F}} =$

$\dfrac{(1.8 \times 10^5 \text{ kg})(0 \text{ m/s} - 15 \text{ m/s})}{-3.5 \times 10^4 \text{ N}}$

$= \boxed{77 \text{ s}}$

$\Delta x = \frac{1}{2}(\mathbf{v_i} + \mathbf{v_f})\Delta t =$

$\frac{1}{2}(15 \text{ m/s} + 0 \text{ m/s})(77 \text{ s}) =$

$\boxed{5.8 \times 10^2 \text{ m}}$

18. 0.33 m/s

Given

$m_1 = 85$ kg

$m_2 = 2.0$ kg

$\mathbf{v_{1,i}} = \mathbf{v_{2,i}} = 0$ m/s

$\mathbf{v_{2,f}} = -14$ m/s

Solution

$m_1\mathbf{v_{1,i}} + m_2\mathbf{v_{2,i}} = m_1\mathbf{v_{1,f}} + m_2\mathbf{v_{2,f}} = 0$

$m_1\mathbf{v_{1,f}} = m_2\mathbf{v_{2,f}}$

$\mathbf{v_{1,f}} = -\dfrac{m_2\mathbf{v_{2,f}}}{m_1} =$

$-\dfrac{(2.0 \text{ kg})(-14 \text{ m/s})}{85 \text{ kg}} = \boxed{0.33 \text{ m/s}}$

19. $= 0.20$ m/s

Given

$m_1 = 0.10$ kg

$m_2 = 0.15$ kg

$\mathbf{v_{2,i}} = 0$ m/s

$\mathbf{v_{1,f}} = -0.045$ m/s

$\mathbf{v_{2,f}} = 0.16$ m/s

Solution

$m_1\mathbf{v_{1,i}} + m_2\mathbf{v_{2,i}} = m_1\mathbf{v_{1,f}} + m_2\mathbf{v_{2,f}}$

$\mathbf{v_{1,i}} = \dfrac{m_1\mathbf{v_{1,f}} + m_2\mathbf{v_{2,f}} - m_2\mathbf{v_{2,i}}}{m_1}$

$\mathbf{v_{1,i}} =$

$\dfrac{(0.10 \text{ kg})(-0.045 \text{ m/s}) + (0.15 \text{ kg})(0.16 \text{ m/s}) - (0.15 \text{ kg})(0 \text{ m/s})}{0.10 \text{ kg}}$

$= \boxed{0.20 \text{ m/s}}$

20. 10 m/s to the north

Given

$m_1 = 90$ kg

$m_2 = 120$ kg

$\mathbf{v_{2,i}} = 4$ m/s to the south $= -4$ m/s

$\mathbf{v_f} = 2$ m/s to the north $= 2$ m/s

$\mathbf{v_{2,f}} = 2.0$ m/s

Solution

$m_1\mathbf{v_{1,i}} + m_2\mathbf{v_{2,i}} = (m_1 + m_2)\mathbf{v_f}$

$\mathbf{v_{1,i}} = -\dfrac{(m_1 + m_2)\mathbf{v_f} - m_2\mathbf{v_{2,i}}}{m_1} =$

$-\dfrac{(90 \text{ kg} + 120 \text{ kg})(2 \text{ m/s}) - (120 \text{ kg})(-4 \text{ m/s})}{90 \text{ kg}} =$

1×10^1 m/s

$\mathbf{v_{1,i}} = \boxed{10 \text{ m/s to the north}}$

Circular Motion and Gravitation

CHAPTER TEST A (GENERAL)

1. c **3.** d

2. c **4.** d

5. c

6. a

Given

$F_1 = 36$ N

$r_2 = 3r_1$

$G = 6.673 \times 10^{-11}$ N•m²/kg²

Solution

$r_2 = 3r_1$

$F_1 = G\dfrac{m_1m_2}{r_1^2} = 36$ N

$F_2 = G\dfrac{m_1m_2}{(r_2)^2} = G\dfrac{m_1m_2}{(3r_1)^2} = G\dfrac{m_1m_2}{9r_1^2}$

$= \dfrac{1}{9}G\dfrac{m_1m_2}{r_1^2} = \dfrac{1}{9}F_1$

$F_2 = \dfrac{1}{9}(36 \text{ N}) = \boxed{4.0 \text{ N}}$

7. d

8. b

9. c

10. a

Solution

$$v_{t_1} = \sqrt{G\frac{m}{r_1}} \qquad v_{t_2} = \sqrt{G\frac{4m}{r_1}}$$

$$\frac{v_{t_2}}{v_{t_1}} = \frac{\sqrt{G\frac{(4m)}{r_1}}}{\sqrt{G\frac{m}{r_1}}} = \sqrt{4} = 2$$

$v_{t_2} = 2v_{t_1}$, i.e., speed would increase by a factor of 2

11. d

12. a

13. b

14. b

Given

$F = 3.0 \times 10^2$ N

$d = 0.80$ m

Solution

$\tau = Fd = (3.0 \times 10^2 \text{ N})(0.80 \text{ m}) = 2.4 \times 10^2$ N•m

15. c

16. c

17. Acceleration depends on the change in an object's velocity. An object moving at a constant speed can experience a nonzero acceleration if the direction of the object's motion changes.

18. Friction between the car's tires and the track provides the centripetal force.

19. No, there is only an inward force causing a deviation from a straight-line path. The tendency to move in a straight line away from the circular path is inertia.

20. A satellite in a circular orbit around Earth moves like a projectile. One component of its motion is parallel to Earth's surface, while the other component is a free-fall acceleration toward Earth. The horizontal component of the motion is just the right magnitude for Earth's curved surface to fall away at the same rate as the satellite's free-fall acceleration.

21. They both accelerate toward each other. Earth's acceleration is extremely small compared to that of the apple because Earth has a much greater mass than the apple does.

22. No, astronauts in orbit are not truly weightless. They experience apparent weightlessness because they are in continual free fall, along with their surrounding environment.

23. 2.1 m/s^2

Given

$v_t = 2.6$ m/s

$r = 3.2$ m

Solution

$$a_c = \frac{v_t^2}{r} = \frac{(2.6 \text{ m/s})^2}{3.2 \text{ m}} = \boxed{2.1 \text{ m/s}^2}$$

24. 74 N

Given

$m = 35$ kg

$v_t = 2.6$ m/s

$r = 3.2$ m

Solution

$$F_c = \frac{mv_t^2}{r} = \frac{(35 \text{ kg})(2.6 \text{ m/s})^2}{3.2 \text{ m}} = \boxed{74 \text{ N}}$$

25. 14.3

Given

$F_{in} = 255$ N

$F_{out} = 3650$ N

Solution

$$MA = \frac{F_{out}}{F_{in}} = \frac{3650 \text{ N}}{255 \text{ N}} = \boxed{14.3}$$

Circular Motion and Gravitation

CHAPTER TEST B (ADVANCED)

1. b

2. d

3. d

Given

$F_1 = 10.0$ N

$r_1 = 10.0$ cm

$r_2 = 5.0$ cm

$G = 6.673 \times 10^{-11}$ N•m^2/kg^2

Solution

$$\frac{r_2}{r_1} = \frac{(5.0 \text{ cm})}{(10.0 \text{ cm})} = \frac{1}{2}$$

$$r_2 = \frac{1}{2}r_1$$

$$F_1 = G\frac{m_1 m_2}{r_1^2} = 10.0 \text{ N}$$

$$F_2 = \frac{G\,m_1 m_2}{(r_2)^2} = G\frac{m_1 m_2}{\left(\frac{1}{2}r_1\right)^2} =$$

$$G\frac{m_1 m_2}{\frac{1}{4}r_1^2} = 4\frac{Gm_1 m_2}{r_1^2} = 4F_1$$

$$F_2 = (4)(10.0\text{ N}) = \boxed{40.0\text{ N}}$$

4. c

Given

$r_2 = 4r_1$

Solution

$$v_{t_1} = \sqrt{G\frac{m}{r_1}} \qquad v_{t_2} = \sqrt{G\frac{m}{4r_1}}$$

$$\frac{v_{t_2}}{v_{t_1}} = \frac{\sqrt{G\dfrac{m}{4r_1}}}{\sqrt{G\dfrac{m}{r_1}}} = \sqrt{\frac{1}{4}} = \frac{1}{2}$$

$v_{t_2} = \frac{1}{2}v_{t_1}$, i.e., speed would decrease by a factor of 2

5. a

6. c

Given

$\tau = 40.0\text{ N}\bullet\text{m}$

$F = 133\text{ N}$

Solution

$\tau = Fd$

$$d = \frac{\tau}{F} = \frac{40.0\text{ N}\bullet\text{m}}{133\text{ N}} = 3.01 \times 10^{-1}\text{ m}$$

$$= \boxed{30.1\text{ cm}}$$

7. b

Given

$F_{in} = 4.0 \times 10^2\text{ N}$

$F_{out} = 6.4 \times 10^3\text{ N}$

Solution

$$MA = \frac{F_{out}}{F_{in}} = \frac{6.4 \times 10^3\text{ N}}{4.0 \times 10^2\text{ N}} = \boxed{16}$$

8. The horse farther from the center has a greater tangential speed. Although both horses complete one circle in the same time period, the one farther from the center covers a greater distance during that time period.

9. Centripetal acceleration: $a_c = \dfrac{v_t^2}{r}$

Newton's second law: $F = ma$

$$F_c = ma_c = \frac{mv_t^2}{r}$$

$$F_c = \frac{mv_t^2}{r}$$

10. 1.0 N

11. Kepler's second law of planetary motion (and, implicitly, Kepler's first law)

12. Newton used Kepler's laws to support his law of universal gravitation. More specifically, he derived Kepler's laws from the law of universal gravitation. This helped support the law of universal gravitation because Kepler's laws closely matched astronomical observations.

13. The object will rotate counterclockwise. It will not move with any translational motion.

14. A machine can increase or decrease the force acting on an object at the expense or gain of the distance moved.

15. 12 m/s^2

Given

$v_t = 17\text{ m/s}$

$r = 24\text{ m}$

Solution

$$a_c = \frac{v_t^2}{r} = \frac{(17\text{ m/s})^2}{24\text{ m}} = \boxed{12\text{ m/s}^2}$$

16. 2.4×10^4 N

Given

$m = 2.0 \times 10^3\text{ kg}$

$v_t = 17\text{ m/s}$

$r = 24\text{ m}$

Solution

$$F_c = \frac{mv_t^2}{r} = \frac{(2.0 \times 10^3\text{ kg})(17\text{ m/s})^2}{24\text{ m}}$$

$$= \boxed{2.4 \times 10^4\text{ N}}$$

17. 9.5×10^3 kg

Given

$m_1 = m_2$

$r = 3.0\text{ m}$

$F_g = 6.7 \times 10^{-4}\text{ N}$

$G = 6.673 \times 10^{-11}\text{ N}\bullet\text{m}^2/\text{kg}^2$

Solution

$m_1 = m_2$

$$F_g = G\frac{m_1 m_2}{r^2} = G\frac{m_1{}^2}{r^2}$$

$$m_1{}^2 = \frac{F_g r^2}{G}$$

$$m_1 = \sqrt{\frac{F_g r^2}{G}} =$$

$$\sqrt{\frac{(6.7 \times 10^{-4}\ \text{N})(3.0\ \text{m})^2}{6.673 \times 10^{-11}\ \text{Nm}^2/\text{kg}^2}} =$$

$$\boxed{9.5 \times 10^3\ \text{kg}}$$

18. Let m_1 be the mass of the central object and m_2 be the mass of the orbiting object.

$$F_g = G\frac{m_1 m_2}{r^2}$$

$$F_c = \frac{m_2 v_t{}^2}{r}$$

The centripetal force equals the gravitational force.

$$F_c = F_g$$

$$\frac{m_2 v_t{}^2}{r} = G\frac{m_1 m_2}{r^2}$$

$$v_t{}^2 = G\frac{m_1}{r}$$

Speed equals the distance traveled in a time interval, and the distance traveled in one orbital period is $2\pi r$.

$$v_t = \frac{2\pi r}{T}$$

Substituting,

$$\left(\frac{2\pi r}{T}\right)^2 = G\frac{m_1}{r}$$

$$\frac{4\pi^2 r^2}{T^2} = G\frac{m_1}{r}$$

$$T^2 = \frac{4\pi^2 r^3}{Gm_1} = \left(\frac{4\pi^2}{Gm_1}\right)r^3$$

Kepler's third law states that $T^2 \propto r^3$.

The constant of proportionality is
$$\frac{4\pi^2}{Gm_1}.$$

19. 7.7×10^4 m; 5.2×10^4 s

Given

$m = 1.0 \times 10^{26}$ kg

$v_t = 9.3 \times 10^3$ m/s

$G = 6.673 \times 10^{-11}$ N•m²/kg²

Solution

$$v_t = \sqrt{G\frac{m}{r}}$$

$$v_t{}^2 = G\frac{m}{r}$$

$$r = G\frac{m}{v_t{}^2} = (6.673 \times 10^{-11}\ \text{N•m}^2/\text{kg}^2)$$

$$\frac{(1.0 \times 10^{26}\ \text{kg})}{(9.3 \times 10^3\ \text{m/s})^2} = \boxed{7.7 \times 10^7\ \text{m}}$$

$$T = 2\pi\sqrt{\frac{r^3}{Gm}} =$$

$$2\pi\sqrt{\frac{(7.7 \times 10^7\ \text{m})^3}{(6.673 \times 10^{-11}\ \text{N•m}^2/\text{kg}^2)(1.0 \times 10^{26}\ \text{kg})}}$$

$$= \boxed{5.2 \times 10^4\ \text{s}}$$

20. 1.0 N•m

Given

$F = 4.0$ N

$d = 0.30$ m

$\theta = 60.0°$

Solution

$$\tau = Fd\sin\theta = (4.0\ \text{N})(0.30\ \text{m})(\sin 60.0°)$$

$$= \boxed{1.0\ \text{N•m}}$$

Fluid Mechanics

CHAPTER TEST A (GENERAL)

1. a   **3.** b

2. b   **4.** c

5. c

Given

$\ell = 10.0$ cm

$\rho_b = 0.780$ g/cm³

$\rho_w = 1.00$ g/cm³

$g = 9.81$ m/s²

Solution

For a floating object,

$$F_B = F_g = mg = \rho Vg = \rho\ell^3 g$$

$$F_B = (0.780\ \text{g/cm}^3)(10.0\ \text{cm})^3$$

$$(9.81\ \text{m/s}^2) \times \left(\frac{1\ \text{kg}}{1000\ \text{g}}\right) = \boxed{7.65\ \text{N}}$$

6. a

7. d

8. c

9. d

Given

$w = 1.5$ m

$\ell = 2.5$ m

$F_g = 1055$ N

Solution

$$P = \frac{F}{A} = \frac{(1055\text{N})}{(1.5\text{ m})(2.5\text{ m})} = \boxed{280 \text{ Pa}}$$

10. a   **13.** d

11. c   **14.** a

12. b   **15.** b

16. c

17. Fluids do not have a definite shape. Solid objects cannot flow, and consequently have a definite shape.

18. The gas expands and changes shape to fill the container.

19. The buoyant force on the object pushes upward on the object so that the net force is less than the weight of the object. The object thus appears to weigh less within the fluid.

20. The net force, or the apparent weight acting on the object, determines whether an object sinks or floats.

21. According to Pascal's principle, the pressure within a fluid is uniform throughout. Therefore, if the pressure on a fluid is known, the pressure throughout the fluid is equal to that known pressure.

22. The pressure in the fluid will decrease.

23. By tapering the hose nozzle, the area of the hose decreases, causing an increase in the speed of the water. This causes the pressure within the water to decrease within the nozzle, and thus increases the pressure difference between the water in the hose and the nozzle. This increased pressure difference pushes the water farther so that it can reach high places that are burning.

24. 7.2×10^{-2} N

Given

$\rho_i = 0.917$ g/cm^3

$\ell = 2.0$ cm

$g = 9.81$ m/s^2

Solution

The ice floats, so

$F_B = F_g = \rho_i V g = \rho_i \ell^3 g$

$F_B = (0.917 \text{ g/cm}^3)(2.0 \text{ cm})^3$

$(9.81 \text{ m/s}^2) \times \left(\dfrac{1 \text{ kg}}{1000 \text{ g}} \right) =$

$\boxed{7.2 \times 10^{-2} \text{ N}}$

25. 530 kg

Given

$A_1 = 0.15$ m^2

$A_2 = 6.0$ m^2

$F_1 = 130$ N

$g = 9.81$ m/s^2

Solution

$P_1 = P_2$

$\dfrac{F_1}{A_1} = \dfrac{F_2}{A_2}$

$m_2 = \dfrac{F_2}{g} = \dfrac{F_1 A_2}{A_1 g} =$

$\dfrac{(130 \text{ N})(6.0 \text{ m}^2)}{(0.15 \text{ m}^2)(9.81 \text{ m/s}^2)} = \boxed{530 \text{ kg}}$

Fluid Mechanics

CHAPTER TEST B (ADVANCED)

1. d

2. d

3. a

Given

$\rho_{Au} = 19.3$ g/cm^3

$m_c = 6.00 \times 10^2$ g

Solution

For a submerged object, the volume of the displaced fluid equals the volume of the object.

$V_w = V_c = \dfrac{m_c}{\rho_{Au}} = \dfrac{6.00 \times 10^2 \text{ g}}{19.3 \text{ g/cm}^3}$

$= \boxed{31.1 \text{ cm}^3}$

4. b

5. c

6. c

Given

$A_{tire} = 0.026$ m^2

$F = 2.6 \times 10^4$ N

number of tires = 4

Solution

The pressure is distributed over the total area provided by 4 tires.

$P = \dfrac{F}{A} = \dfrac{(2.6 \times 10^4 \text{ N})}{(4)(0.026 \text{ m}^2)} =$

$\boxed{2.5 \times 10^5 \text{ Pa}}$

7. a

Given

$F_1 = 230$ N

$F_2 = 6500$ N

$A_1 = 7.0$ m^2

Solution

$P_1 = P_2$

$\dfrac{F_1}{A_1} = \dfrac{F_2}{A_2}$

$A_2 = \dfrac{F_2 A_1}{F_1} = \dfrac{(6500 \text{ N})(7.0 \text{ m}^2)}{(230 \text{ N})}$

$= \boxed{2.0 \times 10^2 \text{ m}^2}$

8. d

9. b

Given

$h_1 = 20.0$ m

$\rho_1 = 1.00$ g/cm^3

$\rho_2 = 13.6$ g/cm^3

Solution

$P = P_0 + \rho g h$

$P - P_0 = \rho_w h_w g = \rho_{Hg} h_{Hg} g$

$h_{Hg} = \dfrac{\rho_w h_w}{\rho_{Hg}} = \dfrac{(1.00 \text{ g/cm}^3)(20.0 \text{ m})}{(13.6 \text{ g/cm}^3)}$

$= \boxed{1.47 \text{ m}}$

10. d

11. The weight of the descending balloon is greater than the buoyant force exerted upward by the air. By reducing the weight of the balloon and gondola, the weight is lowered until it is equal and opposite to the buoyant force, and the balloon remains at a constant elevation.

12. 0.690 g/cm^3, 0.870 g/cm^3, 0.970 g/cm^3, 1.260 g/cm^3

13. 1:4^2, or 1:16

14. The weight of the water behind the dam increases with depth, so the pressure on the dam increases toward the base of the dam. The dam must be thicker at the base to withstand the greater pressure.

15. When there is no water flowing in the tube, the air pressure inside and outside the tube is the same. As the water flows through the tube, pressure within the tube is lowered, causing the air to be pulled inward toward the water.

16. The ball sinks; its apparent weight has a magnitude of 9.6 N.

Given

$\rho_b = 0.940$ g/cm^3

$V_b = 1.4 \times 10^4$ cm^3

$\rho_f = 0.870$ g/cm^3

$g = 9.81$ m/s^2

Solution

$\rho_b > \rho_f$, so the ball sinks.

$F_{net} = F_B - F_g = \rho_f V_f g - \rho_b V_b g$

For the submerged ball, $V_f = V_b$.

$F_{net} = (\rho_f - \rho_b)V_b g$

$F_{net} = (0.870 \text{ g/cm}^3 - 0.940 \text{ g/cm}^3)$
$(1.4 \times 10^4 \text{ cm}^3)(9.81 \text{ m/s}^2) \times$

$\dfrac{1 \text{ kg}}{1000 \text{ g}}$

$F_{net} = (-0.070 \text{ g/cm}^3)$
$(1.4 \times 10^4 \text{ cm}^3)(9.81 \text{ m/s}^2) \times$

$\dfrac{1 \text{ kg}}{1000 \text{ g}}$

$F_{net} = \boxed{-9.6 \text{ N}}$

The ball sinks; its apparent weight has a magnitude of 9.6 N

17. 1.10 g/cm^3

Given

$m = 6.88$ kg

$\rho_w = 1.00$ g/cm^3

apparent weight $= F_{net} = -6.13$ N

$g = 9.81$ m/s^2

Solution

$F_{net} = F_B - F_g = \rho_w V_w g - mg$

$V_w = \dfrac{F_{net} + mg}{\rho_w g} =$

$\dfrac{(-6.13 \text{ N} + (6.88 \text{ kg})(9.81 \text{ m/s}^2))}{(1.00 \text{ g/cm}^3)(9.81 \text{ m/s}^2)} \times$

$\dfrac{1000 \text{ g}}{1 \text{ kg}}$

$V_w = \dfrac{(-6.13 \text{ N} + 67.5 \text{ N})}{(1.00 \text{ g/cm}^3)(9.81 \text{ m/s}^2)} \times$

$\dfrac{1000 \text{ g}}{1 \text{ kg}} = \dfrac{61.4 \text{ N}}{(1.00 \text{ g/cm}^3)(9.81 \text{ m/s}^2)}$

$\times \dfrac{1000 \text{ g}}{1 \text{ kg}}$

$V_w = 6.26 \times 10^3$ cm^3

The weight of the wood is greater than the buoyant force, so the block is submerged, and therefore the volume of the displaced water equals the volume of the block. The density of the wood is equal to its mass over the volume of the displaced water.

$$\rho = \frac{m}{V_w} = \frac{6.88 \text{ kg}}{6.26 \times 10^3 \text{ cm}^3} \times \frac{1000 \text{ g}}{1 \text{ kg}}$$

$$= \boxed{1.10 \text{ g/cm}^3}$$

18. 25.5 m

Given

$P = 3.51 \times 10^5 \text{ Pa}$

$P_0 = 1.01 \times 10^5 \text{ Pa}$

$\rho = 1.00 \times 10^3 \text{ kg/m}^3$

$g = 9.81 \text{ m/s}^2$

Solution

$P = P_0 + \rho g h$

$$h = \frac{P - P_0}{\rho g}$$

$$h = \frac{(3.51 \times 10^5 \text{ Pa} - 1.01 \times 10^5 \text{ Pa})}{(1.00 \times 10^3 \text{ kg/m}^3)(9.81 \text{ m/s}^2)}$$

$$= \frac{2.50 \times 10^5 \text{ Pa}}{(1.00 \times 10^3 \text{ kg/m}^3)(9.81 \text{ m/s}^2)}$$

$h = \boxed{25.5 \text{ m}}$

19. $4.28 \times 10^6 \text{ N}$

Given

$r = 40.0 \text{ cm}$

$h = 850.0 \text{ m}$

$P_0 = 1.01 \times 10^5 \text{ Pa}$

$P_{sub} = 1.31 \times 10^5 \text{ Pa}$

$\rho = 1025 \text{ kg/m}^3$

$g = 9.81 \text{ m/s}^2$

Solution

The pressure exerted on the hatch by the sea water and atmosphere above it is

$P = P_0 + \rho g h$

$P = 1.01 \times 10^5 \text{ Pa} + (1025 \text{ kg/m}^3)$
$(9.81 \text{ m/s}^2)(850.0 \text{ m})$

$P = 1.01 \times 10^5 \text{ Pa} + 8.55 \times 10^6 \text{ Pa} = 8.65 \times 10^6 \text{ Pa}$

The net force on the hatch equals the net pressure multiplied by the area of the hatch.

$F_{net} = (P - P_{sub})A = (P - P_{sub})(\pi r^2)$

$F_{net} =$
$(8.65 \times 10^6 \text{ Pa} - 1.31 \times 10^5 \text{ Pa})$

$(\pi)(40.0 \text{ cm})^2 \times \left(\frac{1 \text{ m}}{100 \text{ cm}}\right)^2$

$F_{net} = (8.52 \times 10^6 \text{ Pa})(\pi)(40.0 \text{ cm})^2 \times$

$\left(\frac{1 \text{ m}}{100 \text{ cm}}\right)^2 = \boxed{4.28 \times 10^6 \text{ N}}$

20. 0.23 m

Given

$v_1 = 15 \text{ m/s}$

$r_1 = 0.40 \text{ m}$

$v_2 = 45 \text{ m/s}$

Solution

$A_1 v_1 = A_2 v_2$

$\pi(r_1)^2 v_1 = \pi(r_2)^2 v_2$

$r_2 = r_1 \sqrt{\frac{v_1}{v_2}} = (0.40 \text{ m})\sqrt{\frac{15 \text{ m/s}}{45 \text{ m/s}}} =$

$\boxed{0.23 \text{ m}}$

Heat

CHAPTER TEST A (GENERAL)

1. b	**8.** c
2. c	**9.** c
3. c	**10.** d
4. c	**11.** a
5. a	**12.** b
6. d	**13.** d
7. a	**14.** b

15. When energy is added to the gas, the kinetic energy of the particles increases. The temperature increases because temperature is proportional to the kinetic energy of the particles.

16. Both objects will have the same final temperature, which will be somewhere between 15° C and 80° C.

17. The temperature of the container is initially greater than the temperature of the air.

18. The energy transfer is the same for both cases, because energy transfer depends on the difference in temperature between the two objects, which is 20° C in both cases.

19. The specific heat capacity of a substance is the amount of heat per unit mass that the substance must absorb to raise the temperature of the substance 1°C (or 1 K) at constant pressure and volume.

20. The ice begins to melt and change into water.

21. The temperature stops rising, and the water turns into steam.

22. The temperature of the melted ice (water) increases steadily until the water begins to vaporize at 100°C.

23. 100.4°F

Given

$T_c = 38.0°$ C

Solution

$$T_F = \frac{9}{5}T_C + 32.0$$

$$T_F = \left(\frac{9}{5}(38.0) + 32.0\right)°F =$$

$$(68.4 + 32.0)°F = \boxed{100.4°F}$$

24. 1.42×10^3 J/kg

Given

$h = 145$ m

$g = 9.81$ m/s^2

Solution

$\Delta PE + \Delta KE + \Delta U = 0$

The kinetic energy increases with the decrease in potential energy, and then decreases with the increase in the internal energy of the water. Thus, the net change in kinetic energy is zero.

$\Delta PE + \Delta U = 0$

Assuming the final potential energy has a value of zero, the change in the internal energy equals

$0 - PE_i + \Delta U = 0$

$\Delta U = PE_i = mgh$

$$\frac{\Delta U}{m} = gh = (9.81 \text{ m/s}^2)(145 \text{ m}) =$$

$$\boxed{1.42 \times 10^3 \text{ J/kg}}$$

25. 29°C

Given

$m = 4.0$ kg

$P = 8.0 \times 10^2$ W

$\Delta t = 10.0$ min

$c_p = 4186$ J/kg•°C

Solution

Heat equals the power delivered multiplied by the time interval.

$Q = P\Delta t$

$Q = c_p m \Delta T$

$$\Delta T = \frac{P\Delta t}{c_p m} = \frac{(8.0 \times 10^2 \text{ W})(10.0 \text{ min})}{(4186 \text{ J/kg°C})(4.0 \text{ kg})}$$

$$\times \left(\frac{60 \text{ s}}{1 \text{ min}}\right) = \boxed{29°C}$$

Heat

CHAPTER TEST B (Advanced)

1. b	**7.** b
2. d	**8.** a
3. c	**9.** b
4. a	**10.** b
5. b	**11.** b
6. c	**12.** c

13. c

14. The volume of many substances, including mercury, increases in proportion to the increase in its temperature. Therefore, by confining mercury to a tube with a constant cross-sectional area, the increase in its volume due to thermal expansion results in a uniform increase in the height of the column, which can be calibrated to given temperatures.

15. Yes, energy is transferred as heat between two objects in thermal equilibrium, but because equal amounts of energy are transferred to and from each object, the net energy transferred is zero, and so the objects remain at their thermal equilibrium temperature.

16. At the microscopic level, energy can be transferred from particles with low kinetic energies (low temperature) to particles with high kinetic energies (high temperature). But this occurs rarely compared to the transfer of energy in collisions from particles with high kinetic energies to those with low kinetic energies. Overall, energy transferred as heat goes from matter at high temperature to matter at low temperature so that objects become cooler, rather than hotter, spontaneously.

17. Mechanical energy is not always conserved. But when the change in internal energy is taken into account along with changes in kinetic and potential energy, the total energy is conserved.

18. 867°F

Given

$T = 737$ K

Solution

$T = T_C + 273.15$

$T_C = T - 273.15$

$$T_F = \frac{9}{5}T_C + 32.0 = \frac{9}{5}(T - 273.15) + 32.0$$

$$T_F = \left(\frac{9}{5}(737 - 273.15) + 32.0\right)°F =$$

$$\left(\frac{9}{5}(464) + 32.0\right)°F = (835 + 32.0)°F$$

$$= \boxed{867°F}$$

19. 979 m

Given

$m = 0.255$ kg

$\Delta U = 2450$ J

$g = 9.81$ m/s^2

Solution

$\Delta PE + \Delta KE + \Delta U = 0$

The kinetic energy increases with the decrease in potential energy, and then decreases with the increase in the internal energy of the water. Thus, the net change in kinetic energy is zero.

$\Delta PE + \Delta U = 0$

Assuming final potential energy has a value of zero, the change in the internal energy equals

$0 - PE_i + \Delta U = 0$

$\Delta U = PE_i = mgh$

$$h = \frac{\Delta U}{mg} = \frac{2450 \text{ J}}{(0.255 \text{ kg})(9.81 \text{ m/s}^2)} =$$

$$\boxed{979 \text{ m}}$$

$$\Delta T = \frac{\frac{1}{4}m_{bullet}v^2}{m_{bullet}\left(\frac{128 \text{ J/kg}}{1.00°C}\right)} =$$

$$\frac{(2.40 \times 10^2 \text{ m/s})^2}{(4)\left(\frac{128 \text{ J/kg}}{1.00°C}\right)} = \boxed{112°C}$$

20. 18°C

Given

$m_{Cu} = 0.10$ kg

$T_{Cu} = 95°C$

$m_w = 0.20$ kg

$m_{Al} = 0.28$ kg

$T_w = T_{Al} = 15°C$

$c_{p,Cu} = 387$ J/kg•°C

$c_{p,Al} = 899$ J/kg•°C

$c_{p,w} = 4186$ J/kg•°C

Solution

From conservation of energy, the energy absorbed as heat by the water and the calorimeter equals the energy given up as heat by the metal.

$Q_w + Q_{Al} = -Q_{Cu}$

$c_{p,w}m_w\Delta T_w + c_{p,Al}m_{Al}\Delta T_{Al} = -c_{p,Cu}m_{Cu}\Delta T_{Cu}$

$c_{p,w}m_w(T_f - T_w) + c_{p,Al}m_{Al}(T_f - T_{Al})$
$= -c_{p,Cu}m_{Cu}(T_f - T_{Cu}) = c_{p,Cu}m_{Cu}(T_{Cu} - T_f)$

$c_{p,w}m_wT_f + c_{p,Al}m_{Al}T_f + c_{p,Cu}m_{Cu}T_f$
$= c_{p,Cu}m_{Cu}T_{Cu} + c_{p,w}m_wT_w + c_{p,Al}m_{Al}T_{Al}$

$$T_f = \frac{c_{p,Cu}m_{Cu}T_{Cu} + c_{p,w}m_wT_w + c_{p,Al}m_{Al}T_{Al}}{c_{p,Cu}m_{Cu} + c_{p,w}m_w + c_{p,Al}m_{Al}}$$

$T_f = [(387$ J/kg•°C$)(0.10$ kg$)(95°C) + (4186$ J/kg•°C$)(0.20$ kg$)(15°C) + (899$ J/kg•°C$)(0.28$ kg$)(15°C)]/ [(387$ J/kg•°C$)(0.10$ kg$) + (4186$ J/kg•°C$)(0.20$ kg$) + (899$ J/kg•°C$)(0.28$ kg$)]$

$$T_f = \frac{(3.7 \times 10^3 \text{ J}) + (1.3 \times 10^4 \text{ J}) + (3.8 \times 10^3 \text{ J})}{(39 \text{ J/°C}) + (8.4 \times 10^2 \text{ J/°C}) + (2.5 \times 10^2 \text{ J/°C})}$$

$$T_f = \frac{2.0 \times 10^4 \text{ J}}{1.13 \times 10^3 \text{ J/°C}} = \boxed{18°C}$$

Thermodynamics

CHAPTER TEST A (GENERAL)

1. b	**8.** b
2. a	**9.** c
3. d	**10.** c
4. c	**11.** b
5. d	**12.** a
6. b	**13.** c
7. a	**14.** c

15. Work is being done on the system (the match and matchbook) to increase the internal energy of the match. When the internal energy (temperature) of the match is high enough for combustion to occur, the chemicals in the match ignite.

16. The process is adiabatic, because no energy is transferred into or out of the system as heat. Work is done on the air in the system, which causes the internal energy of the air to increase.

17. No energy is transferred to or from an isolated system. Therefore, the internal energy of an isolated system remains unchanged.

18. In a cyclic process, the net work equals the net heat.

19. The requirement that $Q_c > 0$ means that some energy must be transferred as heat to the system's surroundings, and therefore this energy cannot be used by the engine to do work.

20. Calculated efficiencies are based only on the amounts of energy transferred as heat to and from the engine. They do not take into account friction or thermal conduction within the engine, which cause energy to be dissipated by the engine. This makes real engines less efficient than their ideal counterparts.

21. Entropy is a measure of the disorder of a system.

22. In most systems, entropy increases with the spontaneous transfer of energy as heat, causing the systems to become more disordered. This process can be reversed, and the system's entropy can be decreased, only by transferring energy as heat from a lower temperature to a higher temperature. This requires work to be done on the system.

23. 5.9×10^5 J

Given

$P = 3.7 \times 10^5$ Pa

$\Delta V = 1.6$ m^3

Solution

$W = P\Delta V = (3.7 \times 10^5 \text{ Pa})(1.6 \text{ m}^3) =$

$\boxed{5.9 \times 10^5 \text{ J}}$

24. -42 J, or 42 J transferred from the system as heat

Given

$W = -165$ J

$\Delta U = 123$ J

Solution

Work is done on the system, so W is negative.

$\Delta U = Q - W$

$Q = \Delta U + W = 123 \text{ J} - 165 \text{ J} =$

-42 J, or 42 J transferred from the system as heat

25. 0.80

Given

$Q_h = 75\,000$ J

$Q_c = 15\,000$ J

Solution

$$eff = 1 - \frac{Q_c}{Q_h}$$

$$eff = 1 - \frac{15\,000 \text{ J}}{75\,000 \text{ J}} = 1 - 0.20 =$$

$\boxed{0.80}$

Thermodynamics

CHAPTER TEST B (ADVANCED)

1. b   **6.** c

2. c   **7.** a

3. d   **8.** d

4. b   **9.** a

5. d   **10.** b

11. Energy from the air was transferred as heat into the balloon. The balloon did work on the book.

12. The volume of the gas decreases.

13. Increasing the net amount of energy transferred as heat from a high-temperature substance to the engine, or decreasing the net amount of energy transferred as heat from the engine to a low-temperature substance, or both of these conditions together will increase the net amount of work done by the engine.

14. Energy is transferred as heat from a high-temperature substance to the lower-temperature engine, and some of the energy is used by the engine to do work on the environment. The remaining energy in the system is transferred as heat from the engine to a lower-temperature substance, which allows work to be done on the engine, thus returning the engine to its initial condition and completing the cycle.

15. According to the second law of thermodynamics, some of the energy added as heat to an engine (Q_h) must be removed from the engine as heat to a substance at a lower temperature (Q_c). Q_c is therefore greater than 0. Efficiency is equal to $1 - (Q_c/Q_h)$, and because Q_c/Q_h must be greater than 0, the efficiency must be less than 1.

16. The entropy of the water decreases, because it goes from a less-ordered liquid state to a more-ordered solid state. This does not occur spontaneously, but by the refrigerator doing work to remove energy as heat from the freezer. This energy is added to the air outside the refrigerator, so the entropy of the outside air (the environment) increases by more than the entropy of the freezing water decreases.

17. -1.4×10^4 J

Given

$P = 7.0 \times 10^4$ N/m^2

$\Delta V = -0.20$ m^3

Solution

The volume decreases, so ΔV, and thus W, are negative.

$W = P\Delta V = (7.0 \times 10^4 \text{ N/m}^2)$

$(-0.20 \text{ m}^3) = \boxed{-1.4 \times 10^4 \text{ J}}$

18. -1.2×10^4 J

Given

$P = 4.13 \times 10^5$ Pa

$r = 0.019$ m

$d = 25.0$ m

$\Delta U = 0$ J

Solution

$\Delta U = Q - W = 0$

Work is done on the system, so W is negative.

$W = P\Delta V = -PAd$

$A = \pi r^2$

$Q = W = -P\pi r^2 d$

$Q = -(4.13 \times 10^5 \text{ Pa})(\pi)(0.019 \text{ m})^2$

$(25.0 \text{ m}) = \boxed{-1.2 \times 10^4 \text{ J}}$

19. 0.257

Given

$Q_h = 2.06 \times 10^{-5}$ J

$Q_c = 1.53 \times 10^{-5}$ J

Solution

$eff = 1 - \dfrac{Q_c}{Q_h}$

$eff = 1 - \dfrac{1.53 \times 10^5 \text{ J}}{2.06 \times 10^5 \text{ J}} = 1 - 0.743 =$

$\boxed{0.257}$

20. 1.69×10^3 J

Given

$\Delta V = 1.50 \times 10^{-3}$ m^3

$P = 3.27 \times 10^5$ Pa

$eff = 0.225$

Solution

$W_{net} = P\Delta V$

$eff = \dfrac{W_{net}}{Q_h}$

$W_{net} = Q_h - Q_c$

$Q_c = Q_h - W_{net} = \dfrac{W_{net}}{eff} - W_{net} =$

$W_{net}\left(\dfrac{1}{eff} - 1\right) = P\Delta V\left(\dfrac{1}{eff} - 1\right)$

$Q_c = (3.27 \times 10^5 \text{ Pa})(1.50 \times 10^{-3} \text{ m}^3)$

$\left(\dfrac{1}{0.225} - 1\right)$

$Q_c = (3.27 \times 10^5 \text{ Pa})(1.50 \times 10^{-3} \text{ m}^3)$

$(4.44 - 1)$

$Q_c = (3.27 \times 10^5 \text{ Pa})(1.50 \times 10^{-3} \text{ m}^3)$

$(3.44) = \boxed{1.69 \times 10^3 \text{ J}}$

Vibrations and Waves

CHAPTER TEST A (GENERAL)

1. a	**10.** a
2. b	**11.** b
3. a	**12.** a
4. d	**13.** c
5. a	**14.** b
6. b	**15.** c
7. c	**16.** c
8. d	**17.** d
9. d	**18.** d

19. three

20. Complete destructive interference should occur because the first pulse is inverted when it reflects from the fixed boundary. The pulses then meet with equal but opposite amplitudes.

21. superposition

22. The period will increase because the restoring force is a component of the gravitational force acting on the pendulum bob (the bob's weight). Because the restoring force is less, but the mass remains the same, the acceleration of the pendulum bob is less.

23. 500 N/m

Given

$F_{elastic} = 50$ N

$x = -0.10$ m

Solution

$F_{elastic} = -kx$

$k = -F_{elastic}/x$

$k = -50 \text{ N}/-0.10 \text{ m} = \boxed{500 \text{ N/m}}$

24. 2.50×10^{-2} Hz

Given

$T = 40.0 \text{ s}$

Solution

$f = 1/T = 1/40.0 \text{ s} = \boxed{2.50 \times 10^{-2} \text{ Hz}}$

25. 40 Hz

Given

$v = 20 \text{ m/s}$

$\lambda = 0.50 \text{ m}$

Solution

$v = f\lambda$

$f = v/\lambda = \dfrac{20 \text{ m/s}}{0.50 \text{ m}} = \boxed{40 \text{ Hz}}$

Vibrations and Waves

CHAPTER TEST B (ADVANCED)

1. b	**8.** d
2. d	**9.** b
3. b	**10.** c
4. b	**11.** a
5. c	**12.** b
6. c	**13.** d
7. c	

14. The restoring force in a swinging pendulum is a component of the gravitational force acting on the pendulum bob.

15. three

16. Moving the pendulum bob down increases the length of the pendulum. As a result, the period of the pendulum increases and the frequency decreases.

17. The water wave is a disturbance moving through the water, but the water (the medium) is not carried forward with the wave.

18. A pulse wave is a single traveling disturbance resulting from a motion that is not repeated. A periodic wave is one whose source is repeated motion.

19. The amplitude increases.

20. (complete) destructive

21. 200 N/m

Given

$x = -0.1 \text{ m}$

$F_{elastic} = 20 \text{ N}$

Solution

$F_{elastic} = -kx$

$k = -F_{elastic}/x$

$k = -20 \text{ N}/-0.1 \text{ m} = \boxed{200 \text{ N/m}}$

22. 1.5 s

Given

$m_{total} = 1500 \text{ kg}$

$k(\text{per spring}) = 6600 \text{ N/m}$

Solution

Assume that the total mass of 1500 kg is supported equally on the four springs. Each spring then supports 1500 kg/4.

$T = 2\pi\sqrt{\dfrac{m}{k}} = 2\pi\sqrt{\dfrac{1500 \text{ kg/4}}{6600 \text{ N/m}}} =$

$\boxed{1.5 \text{ s}}$

23. 21.7 kg

Given

$T_{pendulum} = 3.45 \text{ s}$

$k = 72.0 \text{ N/m}$

Solution

If both systems have the same frequency, they will also have the same period. Therefore, the given period may be substituted into the equation for a mass-spring system.

$T = 2\pi\sqrt{\dfrac{m}{k}}$

$T^2 = 4\pi^2\left(\dfrac{m}{k}\right)$

$m = \dfrac{T^2 k}{4\pi^2} = \dfrac{(3.45 \text{ s})^2(72.0 \text{ N/m})}{4\pi^2} =$

$m = \boxed{21.7 \text{ kg}}$

24. 2.91 m

Given

$f = 103.1 \text{ MHz}$

$v = 3.00 \times 10^8 \text{ m/s}$

Solution

$f = 103.1 \text{ MHz} = 1.031 \times 10^8 \text{ Hz}$

$v = f\lambda$

$\lambda = v/f = (3.00 \times 10^8 \text{ m/s})/$

$(1.031 \times 10^8 \text{ Hz}) = \boxed{2.91 \text{ m}}$

25. 0.80 m

Given

$L = 2.0 \text{ m}$

The standing wave has 5 antinodes, i.e., 5 loops.

Solution

A single loop (antinode) is produced by a wavelength equal to $2L$. Two loops (one complete wavelength) are produced by a wavelength of L. A wavelength of $2/3\ L$ results in 3 antinodes. The following pattern emerges:

1 loop $\lambda = 2L/1 = 2L$
2 loops $\lambda = 2L/2 = L$
3 loops $\lambda = 2L/3 = 2/3\ L$
4 loops $\lambda = 2L/4 = 1/2\ L$
5 loops $\lambda = 2L/5 = 2/5\ L$

$2/5 \times 2.0\ \text{m} = \boxed{0.80\ \text{m}}$

Sound

CHAPTER TEST A (GENERAL)

1. c	11. a
2. d	12. b
3. b	13. c
4. d	14. a
5. a	15. d
6. a	16. a
7. d	17. c
8. a	18. c
9. b	19. b
10. d	20. a

21. compression
22. frequency
23. watts per square meter, or W/m^2
24. Resonance occurs when the frequency of a force applied to an object is the same as the natural frequency of an object.
25. 547 Hz

Given

$v = 684\ \text{m/s}$
$L = 62.5\ \text{cm}$

Solution

$$f_n = n\frac{v}{2L}$$

At the fundamental frequency (first harmonic), $n = 1$, so

$$f_1 = \frac{v}{2L} = \frac{684\ \text{m/s}}{2(0.625\ \text{m})} = \boxed{547\ \text{Hz}}$$

Sound

CHAPTER TEST B (ADVANCED)

1. c	8. c
2. b	9. c
3. c	10. c
4. c	11. d
5. a	12. d
6. c	13. b
7. d	

14. longitudinal
15. rarefaction
16. The pitch rises.
17. As a sphere increases in radius, sections of its surface approach a plane surface. A plane wave is a section of a spherical wave that has such a large radius that sections of it appear planar. This condition appears when the observer of the wave is at a large distance from the source.
18. The apparent pitch of the sound drops as the ambulance passes.
19. The distance from the source doubles, so the intensity decreases to one-fourth of its value at the 10 m distance. The intensity is inversely proportional to the square of the distance from the source, or intensity $\propto 1/r^2$.
20. One of the musical sounds from the CD matches the natural frequency of the string in the piano. As a result, the energy of the sound wave causes the string to vibrate in resonance with the note from the CD.
21. If the musical note is at the fundamental frequency of the glass, the glass will absorb energy from the sound waves and vibrate in resonance with the note. If the sound is loud enough, the vibration will overcome the strength of the glass and the goblet will shatter.
22. $9.3 \times 10^{-3}\ \text{W/m}^2$

Given

$P = 0.30\ \text{W}$
$r = 1.6\ \text{m}$

Solution

$$\text{Intensity} = \frac{P}{4\pi r^2}$$

$$\text{Intensity} = \frac{0.30 \text{ W}}{4\pi (1.6 \text{ m})^2} =$$

$$\boxed{9.3 \times 10^{-3} \text{ W/m}^2}$$

23. 0.350 m

Given

$v = 577$ m/s

$f_1 = 825$ Hz

Solution

$$f_1 = \frac{v}{2L}$$

$$L = \frac{v}{2f_1} = \frac{577 \text{ m/s}}{2(825 \text{ Hz})} = \boxed{0.350 \text{ m}}$$

24. The resonant length must be shortened by 2.5 cm to 25.0 cm.

Given

$v = 348$ m/s

$f_1 = 349$ Hz

$L_{initial} = 27.5$ cm

Solution

$$f_1 = \frac{v}{4L}$$

$$L = \frac{v}{4f_1} = \frac{348 \text{ m/s}}{4(349 \text{ Hz})} = 0.250 \text{ m} =$$

25.0 cm

Change in length $= L_{initial} - L =$

$$27.5 \text{ cm} - 25.0 \text{ cm} = \boxed{2.5 \text{ cm}}$$

25. 565 Hz or 577 Hz

Given

$f_{reference} = 571$ Hz

beats $= 6$/s

Solution

$$f = f_{reference} \pm \text{beats} = 571 \text{ Hz} \pm 6 \text{ Hz}$$

$$f = \boxed{577 \text{ Hz or } 565 \text{ Hz}}$$

Light and Reflection

CHAPTER TEST A (GENERAL)

1. c	**11.** b
2. c	**12.** c
3. a	**13.** d
4. a	**14.** a
5. a	**15.** c
6. a	**16.** a
7. d	**17.** a
8. a	**18.** a
9. d	**19.** d
10. c	**20.** c

21. diffuse

22. through the focal point (F)

23. virtual

24. 0%

25. 5.4×10^{14} Hz

Given

$\lambda = 560$ nm $= 560 \times 10^{-9}$ m $=$

5.6×10^{-7} m

$c = 3.00 \times 10^8$ m/s

Solution

Rearrange the wave speed equation, $c = f\lambda$, to isolate f, and calculate.

$$f = \frac{c}{\lambda} = \frac{(3.00 \times 10^8 \text{ m/s})}{(5.6 \times 10^{-7} \text{ m})} =$$

$$5.4 \times 10^{14} \text{ s}^{-1} = \boxed{5.4 \times 10^{14} \text{ Hz}}$$

Light and Reflection

CHAPTER TEST B (ADVANCED)

1. c

Given

$f = 3.0 \times 10^9$ Hz $= 3.0 \times 10^9$ s^{-1}

$c = 3.00 \times 10^8$ m/s

Solution

Rearrange the wave speed equation, $c = f\lambda$, to isolate λ, and calculate.

$$\lambda = \frac{c}{f} = \frac{(3.00 \times 10^8 \text{ m/s})}{(3.0 \times 10^9 \text{ s}^{-1})} = \boxed{0.10 \text{ m}}$$

2. b

Given

$\lambda = 1.0 \times 10^5$ m

$c = 3.00 \times 10^8$ m/s

Solution

Rearrange the wave speed equation, $c = f\lambda$, to isolate f, and calculate.

$$f = \frac{c}{\lambda} = \frac{(3.00 \times 10^8 \text{ m/s})}{(1.0 \times 10^5 \text{ m})} =$$

$$3.0 \times 10^3 \text{ s}^{-1} = \boxed{3.0 \times 10^3 \text{ Hz}}$$

3. c

4. b

5. b

6. d

7. d

8. a

9. a

10. b

Given

$f = 10.0$ cm

$q = 30.0$ cm

Solution

Rearrange the mirror equation,

$\dfrac{1}{p} + \dfrac{1}{q} = \dfrac{1}{f}$, and solve for p.

$$\dfrac{1}{p} = \dfrac{1}{f} - \dfrac{1}{q} = \dfrac{1}{10.0 \text{ cm}} - \dfrac{1}{30.0 \text{ cm}} =$$

$$\dfrac{3}{30.0 \text{ cm}} - \dfrac{1}{30.0 \text{ cm}} = \dfrac{2}{30.0 \text{ cm}}$$

$$p = \boxed{15 \text{ cm}}$$

11. d

12. d

13. c

14. d

15. The ultraviolet portion of the electromagnetic spectrum is made of sufficiently high frequency (i.e., high energy) electromagnetic radiation that can destroy bacteria or other pathogens.

16. Electromagnetic waves are distinguished by their different frequencies and wavelengths.

17. *Luminous flux* is a measure of the amount of light emitted from a light source. It is measured in lumens. *Illuminance* is a derived unit that indicates the relationship between luminous flux and the distance from the light source squared. Illuminance is the ratio of lumens/m².

18. 52°; According to the law of reflection, the angle of incidence is equal to the angle of reflection.

19. When the candle is at the focal point, the image is infinitely far to the left and therefore is not seen, as shown in the answer diagram.

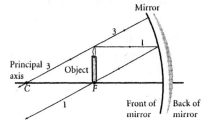

20. $q = -20.0$ cm

$M = +2.00$

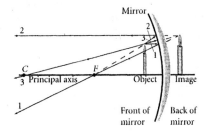

Given

$h = 2.00$ cm

$R = 40.0$ cm

$p = 10.0$ cm

Solution

Since $R = 40.0$ cm, $f = 20.0$ cm.

Rearrange the mirror equation,

$\dfrac{1}{p} + \dfrac{1}{q} = \dfrac{1}{f}$, and solve for q.

$$\dfrac{1}{q} = \dfrac{1}{f} - \dfrac{1}{p} = \dfrac{1}{20.0 \text{ cm}} - \dfrac{1}{10.0 \text{ cm}}$$

$$= \dfrac{1}{20.0 \text{ cm}} - \dfrac{2}{20.0 \text{ cm}} = -\dfrac{1}{20.0 \text{ cm}}$$

$q = -20.0$ cm

Since q is negative, the image is located 2.0×10^1 cm behind the mirror.

$$M = -\dfrac{q}{p} = -\dfrac{(-20.0 \text{ cm})}{10.0 \text{ cm}} = \boxed{+2.00}$$

Refraction

CHAPTER TEST A (GENERAL)

20 mult   3½ pts
4 short   5½ "
1 long   8 "

1. b	**11.** c
2. b	**12.** a
3. c	**13.** a
4. c	**14.** a
5. d	**15.** c
6. a	**16.** b
7. a	**17.** a
8. c	**18.** a
9. d	**19.** b
10. c	**20.** d

21. The speed of light decreases.

22. The image is upright and virtual.

23. The index of refraction of the first medium must be greater than the index of refraction of the second medium.

24. Each colored component of the incoming ray is refracted depending on its wavelength. The rays fan out from the second face of the prism to produce a visible spectrum.

25. 16.7°

Given

$\theta_i = 28.0°$

$n_i = 1.00$

$n_r = 1.63$

Solution

Rearrange Snell's law, $n_i \sin \theta_i = n_r \sin \theta_r$, and solve for θ_r.

$\theta_r = \sin^{-1}\left(\dfrac{n_i}{n_r}(\sin \theta_i)\right) =$

$\sin^{-1}\left(\dfrac{1.00}{1.63}(\sin 28.0°)\right) = \boxed{16.7°}$

Refraction

CHAPTER TEST B (ADVANCED)

1. b

2. c

3. a

4. a

5. b

6. d

Solution

Rearrange Snell's law, $n_i \sin \theta_i = n_r \sin \theta_r$, and solve for θ_r.

$\theta_r = \sin^{-1}\left(\dfrac{n_i}{n_r}(\sin \theta_i)\right) =$

$\sin^{-1}\left(\dfrac{1.00}{1.65}(\sin 3.0 \times 10^{1°})\right) = \boxed{18°}$

7. d

8. a

Solution

Use the thin-lens equation to find f.

$\dfrac{1}{f} = \dfrac{1}{p} + \dfrac{1}{q} = \dfrac{1}{20.0 \text{ cm}} + \dfrac{1}{8.00 \text{ cm}} =$

$\dfrac{0.0500}{1 \text{ cm}} + \dfrac{0.125}{1 \text{ cm}} = \dfrac{0.175}{1 \text{ cm}}$

$f = \boxed{5.71 \text{ cm}}$

9. a

Solution

Use the magnification of a lens equation, $M = \dfrac{h'}{h}$, to find M.

$M = \dfrac{h'}{h} = \dfrac{(151 \text{ cm})}{(1.07 \text{ cm})} = \boxed{141}$

10. c

11. when the difference between the substances' indices of refraction is the greatest

12. An object placed just outside the focal length of the objective lens forms a real, inverted image just inside the focal point of the eyepiece. This eyepiece, the second lens, serves to magnify the image.

13. In order to be seen, the object under a microscope must be at least as large as a wavelength of light. An atom is many times smaller than a wavelength of visible light.

14. A light ray represents the direction of propagation of a planar wave front, which is the superposition of all the spherical wave fronts. As these wave fronts enter a transparent medium, all of them strike the surface simultaneously and experience a similar change in velocity at the same instant. Although this results in a change in the overall wavelength of the spherical wave fronts, there is no change in the direction of the wave fronts relative to each other. Therefore, no refraction occurs.

15. A real, inverted image that is smaller than the object will form between F and $2F$.

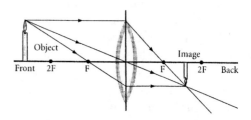

16. The light will undergo total internal reflection.

17. Rays of light from the sun strike Earth's atmosphere and are bent because the atmosphere has an index of refraction greater than that of the near-vacuum of space.

18. Dispersion is the process of separating polychromatic light into its component wavelengths because n is a function of wavelength for all material mediums. Snell's law states that the angles of refraction will be different for different wavelengths even if the angles of incidence are the same.

19. 48 cm

Given

$p = 24$ cm

$f = 16$ cm (f is positive, since this is a converging lens)

Solution

Rearrange the thin-lens equation,

$\dfrac{1}{p} + \dfrac{1}{q} = \dfrac{1}{f}$, and solve for q.

$\dfrac{1}{q} = \dfrac{1}{f} - \dfrac{1}{p} = \dfrac{1}{16 \text{ cm}} - \dfrac{1}{24 \text{ cm}} =$

$\dfrac{0.063}{1 \text{ cm}} - \dfrac{0.042}{1 \text{ cm}} = \dfrac{0.021}{1 \text{ cm}}$

$q = \boxed{48 \text{ cm}}$ (since q is positive, the image is real and in back of the lens)

20. -11 cm

Given

$F_O = 1.00$ cm

$p_O = 1.25$ cm

$F_e = 1.50$ cm

$p_e = 1.50$ cm $- 0.180$ cm $= 1.32$ cm

Solution

The focal length and object distance of the objective lens do not enter into the calculation.

The image of the objective lens is the object of the eyepiece lens.

Rearrange the thin-lens equation,

$\dfrac{1}{p} + \dfrac{1}{q} = \dfrac{1}{f}$, and solve for q.

$\dfrac{1}{q_e} = \dfrac{1}{f_e} - \dfrac{1}{p_e} = \dfrac{1}{1.50 \text{ cm}} - \dfrac{1}{1.32 \text{ cm}} =$

$\dfrac{0.667}{1 \text{ cm}} - \dfrac{0.758}{1 \text{ cm}} = -\dfrac{0.091}{1 \text{ cm}}$

$q_e = \boxed{-11 \text{ cm}}$ (since q is negative, the image is virtual and in front of the lens)

Interference and Diffraction

CHAPTER TEST A (GENERAL)

1. b		**9.** a
2. b		**10.** c
3. a		**11.** b
4. c		**12.** a
5. c		**13.** c
6. d		**14.** d
7. c		**15.** d
8. b		

16. Diffraction is a change in the direction of a wave when the wave encounters an obstacle, an opening, or an edge.

17. spectrometer

18. A spectrometer separates light from a source into its monochromatic components.

19. Resolving power is the ability of an optical instrument to separate two images that are close together.

20. 480 nm

Solution

$d \sin \theta = m\lambda$

$\lambda = \dfrac{d \sin \theta}{m} =$

$\dfrac{(2.5 \times 10^{-6} \text{ m})(\sin 35°)}{3} =$

$4.8 \times 10^{-7} \text{ m} = \boxed{4.8 \times 10^2 \text{ nm}}$

Interference and Diffraction

CHAPTER TEST B (ADVANCED)

1. a	**4.** c
2. b	**5.** c
3. c	**6.** d
7. d	

Solution

$d \sin \theta = m\lambda$

$\lambda = \dfrac{d \sin \theta}{m} =$

$\dfrac{(4.0 \times 10^{-5} \text{ m})(\sin 2.2°)}{2} =$

$7.7 \times 10^{-7} \text{ m} = \boxed{7.7 \times 10^2 \text{ nm}}$

8. d

Solution

$$d \sin \theta = m\lambda$$

$$\theta = \sin^{-1}\left(\frac{m\lambda}{d}\right) =$$

$$\sin^{-1}\left(\frac{3(5.5 \times 10^{-7} \text{ m})}{5.0 \times 10^{-6} \text{ m}}\right) = \boxed{19°}$$

9. c

10. d

Solution

$$d \sin \theta = m\lambda$$

$$\lambda = \frac{d \sin \theta}{m} =$$

$$\frac{\left(\dfrac{1}{5.3 \times 10^3} \text{ cm}\right)(\sin 17°)}{1} =$$

$$5.5 \times 10^{-5} \text{ cm} = \boxed{5.5 \times 10^2 \text{ nm}}$$

11. b

Solution

$$d \sin \theta = m\lambda$$

$$\theta = \sin^{-1}\left(\frac{m\lambda}{d}\right) =$$

$$\sin^{-1}\left(\frac{2(4.000 \times 10^{-7} \text{ m})}{\dfrac{1}{1.00 \times 10^{-6}} \text{ m}}\right) = \boxed{53.1°}$$

12. b

Solution

$$d \sin \theta = m\lambda$$

$$d = \frac{m\lambda}{\sin \theta} = \frac{2.3 \times 10^{-6} \text{ m}}{\sin 27°} =$$

$$5.1 \times 10^{-6} \text{ m} = 5.1 \times 10^{-4} \text{ cm}$$

$$d = 5.1 \times 10^{-4} \text{ cm/line}$$

$$\frac{1}{d} = \frac{1}{5.1 \times 10^{-4} \dfrac{\text{cm}}{\text{line}}} =$$

$$\boxed{2.0 \times 10^3 \text{ lines/cm}}$$

13. The pattern is one of alternating light and dark bands. The brightest light band is at the center and is twice as wide as the other bands. The light bands decrease in brightness as the distance from the center increases.

14. Constantly moving layers of air blur the light from objects in space and limit the resolving power.

15. The resolving power of the instrument will decrease.

16. The waves emitted by a laser do not shift relative to each other as time progresses. The individual waves behave like a single wave because they are coherent and in phase.

17. When energy is added to the active medium, the atoms in the active medium absorb some of the energy. Later, these atoms release energy in the form of light waves that have the equivalent wavelength and phase. The initial waves cause other energized atoms to release their excess energy in the form of more light waves with the same wavelength, phase, and direction as the initial light wave. Mirrors on the end of the material return these coherent light waves into the active medium, where they emit more coherent light waves. One of these mirrors is slightly transparent so that some of the coherent light is emitted.

18. 580 nm

Solution

$$d \sin \theta = m\lambda$$

$$\lambda = \frac{d \sin \theta}{m} =$$

$$\frac{(4.2 \times 10^{-6} \text{ m})(\sin 8.0°)}{1} =$$

$$5.8 \times 10^{-7} \text{ m} = \boxed{5.8 \times 10^2 \text{ nm}}$$

19. 580 nm

Solution

$$d = \frac{1}{6.0 \times 10^3 \dfrac{\text{lines}}{\text{cm}}}$$

$$d \sin \theta = m\lambda$$

$$\lambda = \frac{d \sin \theta}{m} =$$

$$\frac{\left(\dfrac{1}{6.0 \times 10^3} \text{ cm}\right)(\sin 10.0°)}{0.5} =$$

$$5.8 \times 10^{-5} \text{ cm} = \boxed{5.8 \times 10^2 \text{ nm}}$$

20. $\theta_1 = 23.51°$; $\theta_2 = 52.94°$

Solution

$d \sin \theta = m\lambda$

$$\theta_1 = \sin^{-1}\left(\frac{m\lambda}{d}\right) =$$

$$\sin^{-1}\left(\frac{1(6.328 \times 10^{-7} \text{ m})}{\left(\dfrac{1}{6.306\ 92 \times 10^5} \text{ m}\right)}\right) =$$

$$\boxed{23.51°}$$

$$\theta_2 = \sin^{-1}\left(\frac{m\lambda}{d}\right) =$$

$$\sin^{-1}\left(\frac{2(6.328 \times 10^{-7} \text{ m})}{\left(\dfrac{1}{6.306\ 92 \times 10^5} \text{ m}\right)}\right) =$$

$$\boxed{52.94°}$$

Electric Forces and Fields

CHAPTER TEST A (GENERAL) 18 m.ch : 3½ pts
4 short : 5
3 long : 6

1. b	**10.** c
2. b	**11.** b
3. a	**12.** d
4. a	**13.** a
5. b	**14.** d
6. b	**15.** b
7. b	**16.** d
8. c	**17.** b
9. d	**18.** a

19. insulators

20. field

21.

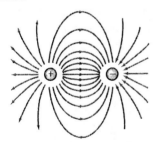

22. field

23. 2.3×10^{-8} N; attractive

Given

$q_e = -e = -1.60 \times 10^{-19}$ C

$q_p = +e = +1.60 \times 10^{-19}$ C

$r = 1.0 \times 10^{-10}$ m

$k_C = 8.99 \times 10^9$ N•m²/C²

Solution

$$F_{electric} = k_C \frac{q_e q_p}{r^2} =$$

$$(8.99 \times 10^9 \text{ N•m}^2/\text{C}^2)$$

$$\left(\frac{(-1.60 \times 10^{-19} \text{ C})(+1.60 \times 10^{-19} \text{ C})}{(1.0 \times 10^{-10} \text{ m})^2}\right)$$

$$F_{electric} = \boxed{-2.3 \times 10^{-8} \text{ N}}$$

24. 1.6×10^{-8} N

Given

$q_e = -e = -1.60 \times 10^{-19}$ C

$q_{nucleus} = +19e = +3.04 \times 10^{-18}$ C

$r = 5.2 \times 10^{-10}$ m

$k_C = 8.99 \times 10^9$ N•m²/C²

Solution

$$F_{electric} = k_C \frac{q_e q_{nucleus}}{r^2} =$$

$$(8.99 \times 10^9 \text{ N•m}^2/\text{C}^2)$$

$$\left(\frac{(-1.60 \times 10^{-19} \text{ C})(+3.04 \times 10^{-18} \text{ C})}{(5.2 \times 10^{-10} \text{ m})^2}\right)$$

$$F_{electric} = \boxed{-1.6 \times 10^{-8} \text{ N}}$$

25. 0.91 m

Given

$r_{A,B} = 2.2$ m

$r_{C,A} = d$

$r_{C,B} = 2.2 \text{ m} - d$

$q_A = 1.0$ C

$q_B = 2.0$ C

$q_C = 2.0$ C

$F_{C,A} = F_{C,B} = 0$ N

Solution

$$F_{C,A} = F_{C,B}$$

$$k_C\left(\frac{q_C q_A}{(r_{C,A})^2}\right) = k_C\left(\frac{q_C q_B}{(r_{C,B})^2}\right)$$

$$\frac{q_A}{d^2} = \frac{q_B}{(2.2 \text{ m} - d)^2}$$

$$(d^2)(q_B) = (2.2 \text{ m} - d)^2(q_A)$$

$$d\sqrt{q_B} = (2.2 \text{ m} - d)\sqrt{q_A}$$

$$d\left(\sqrt{q_B} + \sqrt{q_A}\right) = \sqrt{q_A}(2.2 \text{ m})$$

$$d = \frac{\sqrt{q_A}(2.2 \text{ m})}{\sqrt{q_B} + \sqrt{q_A}} =$$

$$\frac{\sqrt{1.0 \text{ C}}\ (2.2 \text{ m})}{\sqrt{2.0 \text{ C}} + \sqrt{1.0 \text{ C}}} = 0.91 \text{ m}$$

$$d = \boxed{0.91 \text{ m}}$$

Electric Forces and Fields

CHAPTER TEST B (ADVANCED)

1. b   6. c
2. c   7. d
3. a   8. a
4. c   9. d
5. d   10. a

11. Loosely held electrons are transferred from the carpet to the socks when the socks are rubbed against the carpet. The body and socks have an excess of electrons and are negatively charged. Touching the doorknob allows the electrons to escape. The shock felt is the sudden movement of charges as the body and socks return to a neutral state.

12. Millikan discovered that charge is quantized. This means that when any object is charged, the net charge is always a multiple of a fundamental unit of charge. The fundamental unit of charge, which is the charge on the electron, is -1.60×10^{-19} C. The charge on a proton is 1.60×10^{-19} C.

13. The paper becomes charged by polarization. In this process, electrons on each molecule are repelled, and the molecule acquires a positive side near the charged object. As a result, the molecules become attracted to the charged object.

14. The negatively charged rod repels electrons from the part of the sphere nearest the rod. As a result, this part becomes deficient in electrons, thus acquiring a positive charge.

15. because the positive charge is twice the magnitude of the negative charge

16. 1.9×10^{-16} C

    Given

    $q_1 = q_2$
    $F_{electric} = 2.37 \times 10^{-3}$ N
    $r = 3.7 \times 10^{-10}$ m
    $k_C = 8.99 \times 10^9$ N•m^2/C^2

Solution

$$F_{electric} = k_C \frac{q_1 q_2}{r^2} = \frac{k_C q^2}{r^2}$$

$$q = \sqrt{\frac{F_{electric}\, r^2}{k_C}} =$$

$$\sqrt{\frac{(2.37 \times 10^{-3}\ \text{N})(3.7 \times 10^{-10}\ \text{m})^2}{8.99 \times 10^9\ \text{N•m}^2/\text{C}^2}}$$

$$q = \sqrt{\frac{(2.37 \times 10^{-3}\ \text{N})(1.4 \times 10^{-19}\ \text{m}^2)}{8.99 \times 10^9\ \text{N•m}^2/\text{C}^2}}$$

$$q = \boxed{1.9 \times 10^{-16}\ \text{C}}$$

17. $79e$

    Given

    $e = 1.60 \times 10^{-19}$ C
    $q_\alpha = 2e = 3.20 \times 10^{-19}$ C
    $F_{electric} = 91.0$ N
    $r = 2.00 \times 10^{-14}$ m
    $k_C = 8.99 \times 10^9$ N•m^2/C^2

    Solution

$$F_{electric} = k_C \frac{q_\alpha q_{Gold}}{r^2}$$

Rearrange to solve for q_{Gold}.

$$q_{Gold} = \frac{(F_{electric})r^2}{(k_C)q_\alpha} =$$

$$\frac{(91.0\ \text{N})(2.00 \times 10^{-14}\ \text{m})^2}{(8.99 \times 10^9\ \text{N•m}^2/\text{C}^2)(3.20 \times 10^{-19}\ \text{C})} =$$

$$1.27 \times 10^{-17}\ \text{C}$$

$$\frac{q_{Gold}}{q_e} = \frac{1.27 \times 10^{-17}\ \text{C}}{1.60 \times 10^{-19}\ \text{C}} = 79.4$$

The charge on the gold nucleus must be an integer multiple of e.

Integer $(79.4)e = \boxed{79e}$

18. 1.4×10^{-4} N

    Given

    $q_1 = 2.00 \times 10^{-9}$ C
    $q_2 = -3.00 \times 10^{-9}$ C
    $q_3 = 5.00 \times 10^{-9}$ C
    $r_{3,1} = 0.020$ m $= 2.0 \times 10^{-2}$ m
    $r_{3,2} = 0.040$ m $= 4.0 \times 10^{-2}$ m
    $k_C = 8.99 \times 10^9$ N•m^2/C^2

Solution

$$F_{3,1} = k_C \frac{q_3 q_1}{(r_{3,1})^2} =$$

$(8.99 \times 10^9 \text{ N} \bullet \text{m}^2/\text{C}^2)$

$$\frac{(5.00 \times 10^{-9} \text{ C})(2.00 \times 10^{-9} \text{ C})}{(2.0 \times 10^{-2} \text{ m})^2} =$$

$2.2 \times 10^{-4} \text{ N}$

$$F_{3,2} = k_C \frac{q_3 q_2}{(r_{3,2})^2} =$$

$(8.99 \times 10^9 \text{ N} \bullet \text{m}^2/\text{C}^2)$

$$\frac{(5.00 \times 10^{-9} \text{ C})(-3.00 \times 10^{-9} \text{ C})}{(4.0 \times 10^{-2} \text{ m})^2}$$

$= -8.4 \times 10^{-5} \text{ N}$

$F_{tot} = F_{3,1} + F_{3,2} = (2.2 \times 10^{-4} \text{ N}) +$
$(-8.4 \times 10^{-5} \text{ N}) = 1.4 \times 10^{-4} \text{ N}$

$F_3 = \boxed{1.4 \times 10^{-4} \text{ N}}$

19. $1.3 \times 10^9 \text{ N/C}$

Given

$r_1 = r_2 = 2.0 \text{ cm} = 2.0 \times 10^{-2} \text{ m}$
$\theta_1 = 0°$
$\theta_2 = 180°$
$q_1 = 30°\text{C} = 3.0 \times 10^{-5} \text{ C}$
$q_2 = -30°\text{C} = -3.0 \times 10^{-5} \text{ C}$
$k_C = 8.99 \times 10^9 \text{ N} \bullet \text{m}^2/\text{C}^2$

Solution

$$E_1 = k_C \frac{q_1}{r_1^2} = (8.99 \times 10^9 \text{ N} \bullet \text{m}^2/\text{C}^2)$$

$$\left(\frac{3.0 \times 10^{-5} \text{ C}}{(2.0 \times 10^{-2} \text{ m})^2} \right) = 6.7 \times 10^8 \text{ N/C}$$

$$E_2 = k_C \frac{q_2}{r_2^2} = (8.99 \times 10^9 \text{ N} \bullet \text{m}^2/\text{C}^2)$$

$$\left(\frac{-3.0 \times 10^{-5} \text{ C}}{(2.0 \times 10^{-2} \text{ m})^2} \right) = -6.7 \times 10^8 \text{ N/C}$$

For E_1: $E_{x,1} = (E_1)(\cos 0°) =$
$(6.7 \times 10^8 \text{ N/C})(\cos 0°) =$
$6.7 \times 10^8 \text{ N/C}$
$E_{y,1} = 0 \text{ N/C}$
For E_2: $E_{x,2} = (E_2)(\cos 180°) =$
$(-6.7 \times 10^8 \text{ N/C})(\cos 180°) =$
$6.7 \times 10^8 \text{ N/C}$
$E_{y,2} = 0 \text{ N/C}$
$E_{x,tot} = E_{x,1} + E_{x,2} = 6.7 \times 10^8 \text{ N/C} +$
$6.7 \times 10^8 \text{ N/C} = 1.3 \times 10^9 \text{ N/C}$
$E_{y,tot} = E_{y,1} + E_{y,2} =$
$0 \text{ N/C} + 0 \text{ N/C} = 0 \text{ N/C}$

$$E_{tot} = \sqrt{(E_{x,tot})^2 + (E_{y,tot})^2} =$$
$$\sqrt{(1.3 \times 10^9 \text{ N/C})^2 + 0} =$$
$1.3 \times 10^9 \text{ NC}$

$E_{tot} = \boxed{1.3 \times 10^9 \text{ NC}}$

20. $4.8 \times 10^6 \text{ N/C}$

Given

$q_1 = 4.0 \times 10^{-6} \text{ C}$
$q_2 = -6.0 \times 10^{-6} \text{ C}$
$\theta = 60°$
$r_1 = 1.0 \times 10^{-1} \text{ m}$
$r_2 = 1.0 \times 10^{-1} \text{ m}$
$k_C = 8.99 \times 10^9 \text{ N} \bullet \text{m}^2/\text{C}^2$

Solution

$$E_1 = k_C \frac{q_1}{r_1^2} =$$

$(8.99 \times 10^9 \text{ N} \bullet \text{m}^2/\text{C}^2)$

$$\left(\frac{4.0 \times 10^{-6} \text{ C}}{(1.0 \times 10^{-1} \text{ m})^2} \right) = 3.6 \times 10^6 \text{ N/C}$$

$$E_2 = k_C \frac{q_2}{r_2^2} =$$

$(8.99 \times 10^9 \text{ N} \bullet \text{m}^2/\text{C}^2)$

$$\left(\frac{-6.0 \times 10^{-6} \text{ C}}{(1.0 \times 10^{-1} \text{ m})^2} \right) =$$

$-5.4 \times 10^6 \text{ N/C}$
For E_1: $E_{x,1} = (E_1)(\cos 60°) =$
$(3.6 \times 10^6 \text{ N/C})(\cos 60°) =$
$1.8 \times 10^6 \text{ N/C}$
$E_{y,1} = (E_1)(\sin 60°) =$
$(3.6 \times 10^6 \text{ N/C})(\sin 60°) =$
$3.1 \times 10^6 \text{ N/C}$
For E_2: $E_{x,2} = -(E_2)(\cos 60°) =$
$-(-5.4 \times 10^6 \text{ N/C})(\cos 60°) =$
$2.7 \times 10^6 \text{ N/C}$
$E_{y,2} = (E_2)(\sin 60°) =$
$(-5.4 \times 10^6 \text{ N/C})(\sin 60°) =$
$-4.7 \times 10^6 \text{ N/C}$
$E_{x,tot} = E_{x,1} + E_{x,2} =$
$1.8 \times 10^6 \text{ N/C} + 2.7 \times 10^6 \text{ N/C} =$
$4.5 \times 10^6 \text{ N/C}$
$E_{y,tot} = E_{y,1} + E_{y,2} =$
$3.1 \times 10^6 \text{ N/C} + (-4.7 \times 10^6 \text{ N/C}) =$
$-1.6 \times 10^6 \text{ N/C}$

$$E_{tot} = \sqrt{(E_{x,tot})^2 + (E_{y,tot})^2} =$$

$$\sqrt{(4.5 \times 10^6 \text{ N/C})^2 + (-1.6 \times 10^6 \text{ N/C})^2}$$

$= 4.8 \times 10^6 \text{ N/C}$

$E_{tot} = \boxed{4.8 \times 10^6 \text{ N/C}}$

Electrical Energy and Current

CHAPTER TEST A (GENERAL)

1. b **4.** b

2. c **5.** b

3. c

6. b

Solution

$$PE_{electric} = \frac{1}{2}C(\Delta V)^2 =$$

$$\frac{1}{2}(1.5 \times 10^{-6} \text{ F})(9.0 \text{ V})^2 =$$

$$\boxed{6.1 \times 10^{-5} \text{ J}}$$

7. b

8. a

9. b

10. b

Solution

$$\Delta V = IR = (5.0 \text{ A})(5.0 \text{ } \Omega) = \boxed{25 \text{ V}}$$

11. d

12. d

13. a

14. a

Solution

$$P = I\Delta V$$

Rearrange to solve for I.

$$I = \frac{P}{\Delta V} = \frac{75 \text{ W}}{120 \text{ V}} = \boxed{0.62 \text{ A}}$$

15. d

16. Chemical reactions are the source of the energy in a battery.

17. The potential can become so great that a spark discharge or an electrical breakdown will occur.

18. The capacitance is directly proportional to the radius of the sphere.

19. The electron will not move because the wire is connected in an alternating current circuit. The electron moves forward and back the same distance each time the voltage changes direction.

20. The opposition to electric current is resistance.

21. Alternating current is supplied because it is more practical for use in transferring electrical energy.

22. Electric power is the rate at which charge carriers convert electric potential energy to nonelectrical forms of energy.

23. 3.6×10^5 V

Solution

$$r = d = 0.15 \text{ m}$$

$$V = k_C \frac{q}{r} = (8.99 \times 10^9 \text{ N} \bullet \text{m}^2/\text{C}^2)$$

$$\left(\frac{6.0 \times 10^{-6} \text{ C}}{0.15 \text{ m}}\right) = \boxed{3.6 \times 10^5 \text{ V}}$$

24. 2.1 V

Given

$$C = 0.47 \text{ } \mu\text{F} = 4.7 \times 10^{-7} \text{ F}$$

$$Q = 1.0 \text{ } \mu\text{C} = 1.0 \times 10^{-6} \text{ C}$$

Solution

$$C = \frac{Q}{\Delta V}$$

Rearrange to solve for ΔV.

$$\Delta V = \frac{Q}{C} = \frac{1.0 \times 10^{-6} \text{ C}}{4.7 \times 10^{-7} \text{ F}} = \boxed{2.1 \text{ V}}$$

25. 1.1×10^5 A

Given

$$\Delta Q = 9.7 \text{ C}$$

$$\Delta t = 8.9 \times 10^{-5} \text{ s}$$

Solution

$$I = \frac{\Delta Q}{\Delta t} = \frac{9.7 \text{ C}}{8.9 \times 10^{-5} \text{ s}} =$$

$$\boxed{1.1 \times 10^5 \text{ A}}$$

Electrical Energy and Current

CHAPTER TEST B (ADVANCED)

1. b

2. c

3. b

Solution

$$Q = C\Delta V = (0.25 \times 10^{-6} \text{ F})(9.0 \text{ V}) =$$

$$\boxed{2.2 \times 10^{-6} \text{ C}}$$

4. d

Solution

$$PE_{electric} = \frac{1}{2}C(\Delta V)^2 =$$

$$\frac{1}{2}(0.50 \times 10^{-6} \text{ F})(12 \text{ V})^2$$

$$PE_{electric} = \frac{1}{2}(0.50 \times 10^{-6} \text{ F})(144 \text{ V}^2)$$

$$= \boxed{3.6 \times 10^{-5} \text{ J}}$$

5. c

Solution

$$I = \frac{\Delta Q}{\Delta t}$$

Rearrange to solve for ΔQ.

$$\Delta Q = I\Delta t = (7.0 \times 10^{-5} \text{ A})(5.0 \text{ s}) =$$

$$\boxed{3.5 \times 10^{-4} \text{ C}}$$

6. c

7. d

Solution

$\Delta V = IR$

Rearrange to solve for I.

$$I = \frac{\Delta V}{R} = \frac{4.5\ V}{8.0\ \Omega} = \boxed{0.56\ A}$$

8. a

9. b

10. b

Solution

$P = I\Delta V$

Rearrange to solve for ΔV.

$$\Delta V = \frac{P}{I} = \frac{5.00 \times 10^2\ W}{4.00\ A} =$$

$$\boxed{1.25 \times 10^2\ V}$$

11. c

Solution

$P = I\Delta V = I(IR) = I^2 R$

Rearrange to solve for R.

$$R = \frac{P}{I^2} = \frac{325\ W}{(6.0\ A)^2} = \frac{325\ W}{(36.0\ A^2)} =$$

$$\boxed{9.0\ \Omega}$$

12. a

Given

$P = 695\ W$

$\Delta t = 30.0\ min$

Energy cost = \$0.060 per kW•h

Solution

$$\Delta t = (30.0\ min)(\frac{1\ h}{60.0\ min}) = 0.500\ h$$

Energy $= P\Delta t = (695\ W)(0.500\ h) =$
348 W•h

Energy $= (348\ W•h)(\frac{1\ kW}{10^3\ W}) =$
0.348 kW•h

$\$ = (Energy)(Energy\ cost) =$

$$(0.348\ kW•h)(\frac{\$0.060}{1\ kW•h}) = \boxed{\$0.02}$$

13. The electrical potential energy of the charges increases.

14. Increasing the voltage across the capacitor is more effective because the amount of energy stored is proportional to the square of the voltage across the capacitor, while the amount of stored energy is directly proportional to the capacitance of the capacitor.

15. The dielectric reduces the magnitude of the electric field between the plates for a given voltage. The reduced field strength allows the capacitor to operate at a higher voltage for a given plate spacing without causing electrical breakdown.

16. At the instant the bulb is turned on, the lower resistance results in a current through the bulb that is 10 to 20 times greater than the operating current. This high current may cause physical damage to the filament in the light bulb.

17. The power loss at 300 kV is four times the loss at 600 kV. To deliver the same amount of power, the current in the 300 kV line is twice the current in the 600 kV line. The power loss is proportional to the square of the current through the line.

18. 1.7 m

Given

$q = 8.0\ \mu C = 8.0 \times 10^{-6}\ C$

$V = 4.2 \times 10^4\ V$

$k_C = 8.99 \times 10^9\ N•m^2\ /C^2$

Solution

$$V = k_C \frac{q}{r}$$

Rearrange to solve for r.

$$r = k_C \frac{q}{V} = (8.99 \times 10^9\ N•m/C^2)$$

$$\left(\frac{8.0 \times 10^{-6}\ C}{4.2 \times 10^4\ V}\right) = \boxed{1.7\ m}$$

19. $8.3 \times 10^{-5}\ C$

Given

$C = 3.2 \times 10^{-6}\ F$

$\Delta V_1 = 21.0\ V$

$\Delta V_2 = 47.0\ V$

Solution

$Q_1 = C\Delta V_1 = (3.2 \times 10^{-6}\ F)(21.0\ V) =$
$6.7 \times 10^{-5}\ C$

$Q_2 = C\Delta V_2 = (3.2 \times 10^{-6}\ F)(47.0\ V) =$
$1.5 \times 10^{-4}\ C$

$Q = Q_2 - Q_1 =$
$(1.5 \times 10^{-4}\ C) - (6.7 \times 10^{-5}\ C)$

$Q = \boxed{8.3 \times 10^{-5}\ C}$

20. 16 Ω

Given

$\Delta V = 123$ V

$P = 0.95$ kW

Solution

$$P = 0.95 \text{ kW}\left(\frac{1.0 \times 10^3 \text{ W}}{1 \text{ kW}}\right) =$$

9.5×10^2 W

$$P = \frac{(\Delta V)^2}{R}$$

Rearrange to solve for R.

$$R = \frac{(\Delta V)^2}{P} = \frac{(123 \text{ V})^2}{9.5 \times 10^2 \text{ W}} =$$

$$\frac{15129 \text{ V}^2}{9.5 \times 10^2 \text{ W}} = \boxed{16 \text{ } \Omega}$$

Circuits and Circuit Elements

CHAPTER TEST A (GENERAL)

1. c

2. b

3. c

4. c

5. a

6. a

7. a

8. c

9. d

10. b

Given

$R_1 = 3.0$ Ω

$R_2 = 6.0$ Ω

$R_3 = 12$ Ω

Solution

$$\frac{1}{R_{eq}} = \frac{1}{R_1} + \frac{1}{R_2} + \frac{1}{R_3} = \frac{1}{3.0 \text{ } \Omega} +$$

$$\frac{1}{6.0 \text{ } \Omega} + \frac{1}{12 \text{ } \Omega}$$

$$\frac{1}{R_{eq}} = \frac{4.0}{12 \text{ } \Omega} + \frac{2.0}{12 \text{ } \Omega} + \frac{1.0}{12 \text{ } \Omega} =$$

$$\frac{7.0}{12 \text{ } \Omega}$$

$$R_{eq} = \frac{12 \text{ } \Omega}{7.0} = \boxed{1.7 \text{ } \Omega}$$

11. b

12. c

13. c

Given

$R_1 = 6.0$ Ω

$R_2 = 12$ Ω

$R_3 = 4.0$ Ω

Solution

$$\frac{1}{R_{1,2}} = \frac{1}{R_1} + \frac{1}{R_2} = \frac{1}{6.0 \text{ } \Omega} + \frac{1}{12 \text{ } \Omega}$$

$$\frac{1}{R_{1,2}} = \frac{2.0}{12 \text{ } \Omega} + \frac{1.0}{12 \text{ } \Omega} = \frac{3.0}{12 \text{ } \Omega}$$

$$R_{1,2} = \frac{12 \text{ } \Omega}{3.0} = 4.0 \text{ } \Omega$$

$$R_{1,2,3} = R_{1,2} + R_3 = 4.0 \text{ } \Omega + 4.0 \text{ } \Omega =$$

$$\boxed{8.0 \text{ } \Omega}$$

14. b

Given

$R_1 = 3.0$ Ω

$R_2 = 3.0$ Ω

$R_3 = 3.0$ Ω

$R_4 = 3.0$ Ω

Solution

$$R_{1,2,3} = R_1 + R_2 + R_3 =$$

$$3.0 \text{ } \Omega + 3.0 \text{ } \Omega + 3.0 \text{ } \Omega = 9.0 \text{ } \Omega$$

$$\frac{1}{R_{eq}} = \frac{1}{R_{1,2,3}} + \frac{1}{R_4} =$$

$$\frac{1}{9.0 \text{ } \Omega} + \frac{1}{3.0 \text{ } \Omega}$$

$$\frac{1}{R_{eq}} = \frac{1.0}{9.0 \text{ } \Omega} + \frac{3.0}{9.0 \text{ } \Omega} = \frac{4.0}{9.0 \text{ } \Omega}$$

$$R_{eq} = \frac{9.0 \text{ } \Omega}{4.0} = \boxed{2.2 \text{ } \Omega}$$

15. b

16. two batteries, three resistors

17. Bulb A has a current, but B and C do not because the switch is open.

18. A battery has a small internal resistance. As the current increases, the potential difference across this internal resistance increases and reduces the potential difference measured at the terminals of the battery.

19. 27 Ω

Given

$I_{R1} = 0.20$ A

$R_1 = 3.0$ Ω

$\Delta V_{batt} = 6.0$ V

Solution

$\Delta V_{R1} = R_1 \times I_1 = 3.0 \text{ } \Omega \times 0.20 \text{ A} = $
0.60 V

$\Delta V_{R2} = \Delta V_{batt} - \Delta V_{R1} = $
6.0 V − 0.60 V = 5.4 V

$I_{R2} = I_{R1} = 0.20$ A

$$R_2 = \frac{V_{R2}}{I_{R2}} = \frac{5.4 \text{ V}}{0.20 \text{ A}} = \boxed{27 \text{ } \Omega}$$

20. $9.0\ \Omega$

Given

$R_1 = 27\ \Omega$
$R_2 = 81\ \Omega$
$R_3 = 16\ \Omega$

Solution

$$\frac{1}{R_{eq}} = \frac{1}{R_1} + \frac{1}{R_2} + \frac{1}{R_3} =$$

$$\frac{1}{27\ \Omega} + \frac{1}{81\ \Omega} + \frac{1}{16\ \Omega}$$

$$\frac{1}{R_{eq}} = \frac{0.037}{1\ \Omega} + \frac{0.012}{1\ \Omega} + \frac{0.062}{1\ \Omega} =$$

$$\frac{0.111}{1\ \Omega}$$

$$\frac{1}{R_{eq}} = \frac{1\ \Omega}{0.111} = \boxed{9.0\ \Omega}$$

Circuits and Circuit Elements

CHAPTER TEST B (ADVANCED)

1. d
2. b
3. c
4. a

Given

$R_1 = 4.0\ \Omega$
$R_2 = 6.0\ \Omega$
$R_3 = 8.0\ \Omega$

Solution

$R_{eq} = R_1 + R_2 + R_3 =$
$\quad 4.0\ \Omega + 6.0\ \Omega + 8.0\ \Omega = 18\ \Omega$

5. c
6. c
7. d

Given

$R_1 = 4.0\ \Omega$
$R_2 = 6.0\ \Omega$
$R_3 = 10.0\ \Omega$

Solution

$$\frac{1}{R_{eq}} = \frac{1}{R_1} + \frac{1}{R_2} + \frac{1}{R_3} =$$

$$\frac{1}{4.0\ \Omega} + \frac{1}{6.0\ \Omega} + \frac{1}{10.0\ \Omega}$$

$$\frac{1}{R_{eq}} = \frac{0.25}{1\ \Omega} + \frac{0.17}{1\ \Omega} + \frac{0.100}{1\ \Omega} =$$

$$\frac{0.520}{1\ \Omega}$$

$$R_{eq} = \frac{1\ \Omega}{0.520} = \boxed{1.9\ \Omega}$$

8. b

Given

$R_1 = 10.0\ \Omega$
$R_2 = 10.0\ \Omega$
$R_3 = 16\ \Omega$
$R_4 = 8.0\ \Omega$
$R_5 = 8.0\ \Omega$
$\Delta V = 60\ \text{V}$

Solution

$R_{1,2} = R_1 + R_2 =$
$\quad 10.0\ \Omega + 10.0\ \Omega = 20.0\ \Omega$

$$\frac{1}{R_{4,5}} = \frac{1}{R_4} + \frac{1}{R_5} = \frac{1}{8.0\ \Omega} + \frac{1}{8.0\ \Omega}$$

$$\frac{1}{R_{4,5}} = \frac{2.0}{8.0\ \Omega}$$

$$R_{4,5} = \frac{8.0\ \Omega}{2.0} = 4.0\ \Omega$$

$R_{3,4,5} = R_3 + R_{4,5} =$
$\quad 16\ \Omega + 4.0\ \Omega = 20\ \Omega$

$$\frac{1}{R_{eq}} = \frac{1}{R_{1,2}} + \frac{1}{R_{3,4,5}} =$$

$$\frac{1}{20.0\ \Omega} + \frac{1}{20\ \Omega}$$

$$\frac{1}{R_{eq}} = \frac{2.0}{20.0\ \Omega}$$

$$R_{eq} = \frac{20.0\ \Omega}{2} = \boxed{10.0\ \Omega}$$

9. b

Given

$R_1 = 8.0\ \Omega$
$R_2 = 2.0\ \Omega$
$R_3 = 10.0\ \Omega$
$R_4 = 5.0\ \Omega$

Solution

$R_{1,2} = R_1 + R_2 = 8.0\ \Omega + 2.0\ \Omega =$
$\quad 10.0\ \Omega$

$$\frac{1}{R_{1,2,3}} = \frac{1}{R_{1,2}} + \frac{1}{R_3} =$$

$$\frac{1}{10.0\ \Omega} + \frac{1}{10.0\ \Omega} = \frac{2}{10.0\ \Omega}$$

$$R_{1,2,3} = \frac{10.0\ \Omega}{2} = 5.00\ \Omega$$

$R_{eq} = R_{1,2,3} + R_4 =$
$\quad 5.00\ \Omega + 5.0\ \Omega = \boxed{10.0\ \Omega}$

10. a

Given

$R_1 = 2.0\ \Omega$

$R_2 = 4.0\ \Omega$

$R_3 = 6.0\ \Omega$

$R_4 = 10.0\ \Omega$

$\Delta V_{batt} = 12\ V$

Solution

$$\frac{1}{R_{2,3,4}} = \frac{1}{R_2} + \frac{1}{R_3} + \frac{1}{R_4} =$$

$$\frac{1}{4.0\ \Omega} + \frac{1}{6.0\ \Omega} + \frac{1}{10.0\ \Omega}$$

$$\frac{1}{R_{2,3,4}} = \frac{0.25}{1\ \Omega} + \frac{0.17}{1\ \Omega} + \frac{0.100}{1\ \Omega} =$$

$$\frac{0.520}{1\ \Omega}$$

$$R_{2,3,4} = \frac{1\ \Omega}{0.520} = 1.92\ \Omega$$

$$R_{1,2,3,4} = R_1 + R_{2,3,4} =$$
$$2.0\ \Omega + 1.92\ \Omega = 3.9\ \Omega$$

$$I_{total} = \frac{\Delta V_{batt}}{R_{1,2,3,4}} = \frac{12\ V}{3.9\ \Omega} = 3.1\ A$$

$$\Delta V_{R1} = R_1 \times I_{total} =$$
$$2.0\ \Omega \times 3.1\ A = 6.2\ V$$

$$\Delta V_{R4} = \Delta V_{batt} - \Delta V_{R1} =$$
$$12\ V - 6.2\ V = 5.8\ V$$

$$I_{R4} = \frac{\Delta V_{R4}}{R_4} = \frac{5.8\ V}{10.0\ \Omega} = \boxed{0.58\ A}$$

11. Schematics should show all the named circuit elements wired in series, but the order is not important.

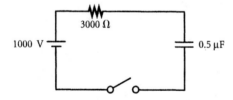

1000 V | 3000 Ω | 0.5 μF

12. the load

13. The lamp will go out, or the lamp will get dimmer. The resistance through the switch is lower than the filament of the lamp.

14. The current in the circuit increases because the equivalent resistance is lower than it was before the replacement.

15. Connect five of the resistors in parallel to produce a 20.0 Ω equivalent resistance. Connect the remaining two resistors in parallel to produce a 50.0 Ω equivalent resistance. Then connect the two groups in series, giving an equivalent resistance of 70.0 Ω.

16. The voltage across R_a is equal to the voltage across R_d, and the voltage across R_c is equal to the voltage across R_f. The voltage across one pair is not necessarily equal to the voltage across the other pair.

17. A high-resistance voltmeter would not alter the resistance of the circuit as much as a voltmeter with a lower resistance. Because the resistance of the voltmeter is in parallel with the resistor across which the voltage is being measured, a high-resistance voltmeter will give a more accurate voltage reading than a voltmeter with a lower resistance.

18. 27 Ω

Given

$I_{R1} = 0.20\ A$

$R_1 = 3.0\ \Omega$

$\Delta V_{batt} = 6.0\ V$

Solution

$$\Delta V_{R1} = R_1 \times I_1 = 3.0\ \Omega \times 0.20\ A = 0.60\ V$$

$$\Delta V_{R2} = \Delta V_{batt} - \Delta V_{R1} = 6.0\ V - 0.60\ V = 5.4\ V$$

$$I_{R2} = I_{R1} = 0.20\ A$$

$$R_2 = \frac{V_{R2}}{I_{R2}} = \frac{5.4\ V}{0.20} = \boxed{27\ \Omega}$$

19. 9.3 Ω

Given

$R_1 = 15\ \Omega$

$R_2 = 41\ \Omega$

$R_3 = 58\ \Omega$

Solution

$$\frac{1}{R_{eq}} = \frac{1}{R_1} + \frac{1}{R_2} + \frac{1}{R_3} =$$

$$\frac{1}{15\ \Omega} + \frac{1}{41\ \Omega} + \frac{1}{58\ \Omega}$$

$$\frac{1}{R_{eq}} = \frac{0.067}{1\ \Omega} + \frac{0.024}{1\ \Omega} + \frac{0.017}{1\ \Omega} =$$

$$\frac{0.108}{1\ \Omega}$$

$$R_{eq} = \frac{1\ \Omega}{0.108} = \boxed{9.3\ \Omega}$$

20. 0.80 A

Given

$R_1 = 2.0\ \Omega$
$R_2 = 20.0\ \Omega$
$R_3 = 10.0\ \Omega$
$R_4 = 10.0\ \Omega$
$\Delta V_{batt} = 12\ \text{V}$
$I_{R3} = I_{R4}$

Solution

$$\frac{1}{R_{2,3,4}} = \frac{1}{R_2} + \frac{1}{R_3} + \frac{1}{R_4} = \frac{1}{20.0\ \Omega} +$$

$$\frac{1}{10.0\ \Omega} + \frac{1}{10.0\ \Omega}$$

$$\frac{1}{R_{2,3,4}} = \frac{1}{20.0\ \Omega} + \frac{2}{20.0\ \Omega} + \frac{2}{20.0\ \Omega} =$$

$$\frac{5}{20.0\ \Omega}$$

$$R_{2,3,4} = \frac{20.0\ \Omega}{5} = 4.00\ \Omega$$

$$R_{eq} = R_1 + R_{2,3,4} =$$
$$2.0\ \Omega + 4.00\ \Omega = 6.0\ \Omega$$

$$I_{tot} = \frac{\Delta V_{batt}}{R_{eq}} = \frac{12\ \text{V}}{6.0\ \Omega} = 2.0\ \text{A}$$

$$\Delta V_3 = \Delta V_4 = R_{2,3,4} \times I_{tot} =$$
$$4.00\ \Omega \times 2.0\ \text{A} = 8.0\ \text{V}$$

$$I_{R3} = I_{R4} = \frac{\Delta V_3}{R_3} = \frac{\Delta V_4}{R_4} =$$

$$\frac{8.0\ \text{V}}{10.0\ \Omega} = \boxed{0.80\ \text{A}}$$

Magnetism

CHAPTER TEST A (GENERAL)

1. b	**9.** b
2. a	**10.** c
3. d	**11.** c
4. a	**12.** b
5. c	**13.** a
6. a	**14.** d
7. d	**15.** b
8. d	**16.** a

17. Magnetic poles cannot be isolated no matter how many times a magnet is cut or subdivided.

18. repel

19. attract

20. $3.6 \times 10^{-4}\ \text{T}$

Given

$v = 9.8 \times 10^4\ \text{m/s, north}$
$q_{electron} = 1.60 \times 10^{-19}\ \text{C}$
$F_{magnetic} = 5.6 \times 10^{-18}\ \text{N, west}$

Solution

$$B = \frac{F_{magnetic}}{q_{electron}v} =$$

$$\frac{(5.6 \times 10^{-18}\ \text{N})}{(1.60 \times 10^{-19}\ \text{C})(9.8 \times 10^4\ \text{m/s})} =$$

$$\boxed{3.6 \times 10^{-4}\ \text{T}}$$

Magnetism

CHAPTER TEST B (ADVANCED)

1. a

2. b

3. d

4. d

5. a

Given

$v = 4.5 \times 10^4\ \text{m/s}$
$q_{electron} = 1.60 \times 10^{-19}\ \text{C}$
$F_{magnetic} = 7.2 \times 10^{-18}\ \text{N}$

Solution

$$B = \frac{F_{magnetic}}{q_{electron}v} =$$

$$\frac{(7.2 \times 10^{-18}\ \text{N})}{(1.60 \times 10^{-19}\ \text{C})(4.5 \times 10^4\ \text{m/s})} =$$

$$1.0 \times 10^{-3}\ \text{T or } \boxed{1.0\ \text{mT}}$$

6. b

Given

$v = 3.0 \times 10^4$ m/s

$q_{electron} = 1.60 \times 10^{-19}$ C

$B = 0.40$ T

Solution

Rearrange equation, $B = \dfrac{F_{magnetic}}{qv}$,

and solve for $F_{magnetic}$.

$F_{magnetic} = q_{electron}vB =$
$(1.60 \times 10^{-19}$ C$)(3.0 \times 10^4$ m/s$)$
$(0.40$ T$) = \boxed{1.9 \times 10^{-15}\ \text{N}}$

7. c

Given

$v = 2.5 \times 10^6$ m/s

$q_{electron} = 1.60 \times 10^{-19}$ C

$B = 0.10 \times 10^{-4}$ T

$\theta = 35°$ north of east

Solution

Since only the magnetic field component that is perpendicular to the electron's motion contributes to the magnetic field strength, $B_{net} = B\cos\theta$. Therefore, $B_{net} = (0.10 \times 10^{-4}$ T$)$
$(\cos 35°) = 8.2 \times 10^{-6}$ T.

Rearrange equation, $B = \dfrac{F_{magnetic}}{qv}$,

and solve for $F_{magnetic}$.

$F_{magnetic} = q_{electron}vB =$
$(1.60 \times 10^{-19}$ C$)(2.5 \times 10^6$ m/s$)$
$(8.2 \times 10^{-6}$ T$) = \boxed{3.3 \times 10^{-18}\ \text{N}}$

8. c

9. d

Given

$B = 0.50$ T

$I = 0.60$ A

$l = 2.0$ m

Solution

Since the wire is oriented parallel to the magnetic field, $B_{net} = 0.0$ T.

$F_{magnetic} = BIl = (0.0$ T$)(0.60$ A$)$
$(2.0$ m$) = \boxed{0.0\ \text{N}}$

10. a

11. the south pole

12. The end of the magnetized nail touching the north pole of the magnet will be a south pole by induction. Otherwise, the magnet would repel it. The tip of the nail that points away from the magnet must have the opposite polarity and thus will be a north pole.

13. According to the magnetic domain model, both hard and soft magnetic materials have large groups of neighboring atoms whose net electron spins are aligned. Soft magnetic materials tend to gain or lose their domain alignments easily, while hard magnetic materials retain their domain alignments. As a result, soft magnetic materials are easily magnetized, but also lose their magnetism easily. Hard magnetic materials are difficult to magnetize, but keep their magnetism for long periods of time.

14. A

15. With the thumb in the direction of the current, the fingers will curl down around the loop. Thus the magnetic field points downward around the loop. The north pole of the loop is below the loop, since the magnetic field appears to be exiting the loop area.

16. solenoid; left end, since the magnetic field lines are entering the solenoid at this end; Using the right-hand rule, the current is entering the wire at the top right.

17. up, toward the top of the page

18. 1.7×10^{-4} T; Using the right-hand rule, $F_{magnetic}$ is upward, since the charged particle is an electron.

Given

$v = 7.3 \times 10^4$ m/s

$\theta = 25°$ south of east

$q_{electron} = 1.60 \times 10^{-19}$ C

$F_{magnetic} = 1.8 \times 10^{-18}$ N

B direction is south.

Solution
Since only the magnetic field component that is perpendicular to the electron's motion contributes to the magnetic field strength, $B_{net} = B\cos\theta$. Substituting $B\cos\theta$ into the equation,

$B = \dfrac{F_{magnetic}}{qv}$, results in $B\cos\theta =$

$\dfrac{F_{magnetic}}{qv}$.

Rearrange equation, $B\cos\theta =$

$\dfrac{F_{magnetic}}{qv}$, and solve for B.

$B = \dfrac{F_{magnetic}}{q_{electron}v\cos\theta} =$

$\dfrac{(1.8 \times 10^{-18}\text{ N})}{(1.60 \times 10^{-19}\text{ C})(7.3 \times 10^4\text{ m/s})(\cos 25°)} =$

$\boxed{1.7 \times 10^{-4}\text{ T}}$

Using the right-hand rule, $F_{magnetic}$ is upward, since the charged particle is an electron.

19. 7.24×10^{-8} s; 6.19×10^6 m/s

Given
$B = 6.48 \times 10^{-2}$ T
$F_{magnetic} = 7.16 \times 10^{-14}$ N
$q_{proton} = 1.60 \times 10^{-19}$ C
$\Delta x = 0.500$ m

Solution

$v = \dfrac{\Delta x}{\Delta t}$

Substitute for v in the equation, $B =$

$\dfrac{F_{magnetic}}{qv} = \dfrac{F_{magnetic}}{q\left(\dfrac{\Delta x}{\Delta t}\right)}$

Rearrange the equation and solve for Δt.

$\Delta t = \dfrac{Bq_{proton}\Delta x}{F_{magnetic}} =$

$\dfrac{(6.48 \times 10^{-2}\text{ T})(1.60 \times 10^{-19}\text{ C})(0.500\text{ m})}{(7.16 \times 10^{-14}\text{ N})}$

$= \boxed{7.24 \times 10^{-8}\text{ s}}$

$v = \dfrac{\Delta x}{\Delta t} = \dfrac{(0.500\text{ m})}{(7.24 \times 10^{-8}\text{ s})} =$

$\boxed{6.19 \times 10^6\text{ m/s}}$

20. 7.2×10^{-1} N, downward
Given
$B = 8.3 \times 10^{-4}$ T
$I = 18$ A
$l = 48$ m

Solution
$F_{magnetic} = BIl =$
$\quad (8.3 \times 10^{-4}\text{ T})(18\text{ A})(48\text{ m}) =$
$\boxed{7.2 \times 10^{-1}\text{ N, downward}}$

Electromagnetic Induction

CHAPTER TEST A (GENERAL)

1. c	**6.** d
2. d	**7.** b
3. b	**8.** c
4. d	**9.** c
5. a	**10.** d

11. c
Given
$\Delta V_{max} = 220$ V

Solution
$\Delta V_{rms} = 0.707\Delta V_{max} =$
$\quad (0.707)(220\text{ V}) = \boxed{160\text{ V}}$

12. a
Given
$I_{rms} = 3.6$ A

Solution
Rearrange the equation, $I_{rms} = 0.707 I_{max} =$, to solve for I_{max}.

$I_{max} = \dfrac{I_{rms}}{0.707} = \dfrac{(3.6\text{ A})}{(0.707)} = \boxed{5.1\text{ A}}$

13. d
Given
$\Delta V_1 = 120$ V
$N_1 = 95$ turns
$N_2 = 2850$ turns

Solution

$\Delta V_2 = \dfrac{\Delta V_1 N_2}{N_1} = (120\text{ V})\left(\dfrac{2850\text{ turns}}{95\text{ turns}}\right)$

$= \boxed{3600\text{ V}}$

14. b
Given
$\Delta V_1 = 115$ V
$\Delta V_2 = 2.3$ V

Solution

$N_1{:}N_2 = \dfrac{\Delta V_1}{\Delta V_2} = \dfrac{115\text{ V}}{2.3\text{ V}} = \boxed{50{:}1}$

15. c
16. b
17. b
18. b
19. Move the circuit loop into or out of the magnetic field. Rotate the circuit loop in the magnetic field so that the angle between the plane of the circuit loop and magnetic field changes. Vary the intensity of the magnetic field by rotating the magnet.
20. to turn wire loops (or a coil of wire) in a magnetic field
21. the rms current multiplied by the resistance (or $I_{rms} \bullet R$)
22. The energy of an electromagnetic wave is stored in the electromagnetic fields, which exert an electromagnetic force on charged particles. This energy transported by an electromagnetic wave is called electromagnetic radiation.
23. -149 V

Given
$N = 275$ turns
$A = 0.750$ m^2
$\Delta B = +0.900$ T
$\Delta t = 1.25$ s
$\theta = 0.00°$

Solution
Substitute values into Faraday's law of magnetic induction.

$$\text{emf} = \frac{-N\Delta\Phi_M}{\Delta t} = -N\frac{\Delta AB\cos\theta}{\Delta t} =$$

$$-NA\cos\theta\frac{\Delta B}{\Delta t}$$

$$= -(275 \text{ turns})(0.750 \text{ m}^2)$$

$$(\cos 0.00°)\frac{(0.900 \text{ T})}{(1.25 \text{ s})} = \boxed{-149 \text{ V}}$$

24. 297 V

Given
$\Delta V_{max} = 4.20 \times 10^2$ V

Solution

$\Delta V_{rms} \doteq 0.707\Delta V_{max} =$
$(0.707)(4.2 \times 10^2 \text{ V}) = \boxed{297 \text{ V}}$

25. 1.6×10^4 V

Given
$\Delta V_1 = 120$ V
$N_1 = 38$ turns
$N_2 = 5163$ turns

Solution

$$\Delta V_2 = \Delta V_1\frac{N_2}{N_1} = (120 \text{ V})\left(\frac{5163 \text{ turns}}{38 \text{ turns}}\right)$$

$$= \boxed{1.6 \times 10^4 \text{ V}}$$

Electromagnetic Induction

CHAPTER TEST B (ADVANCED)

1. c
2. a
3. b
4. a
5. a
6. d
7. a
8. b

Given
$\Delta V_{max} = 215$ V

Solution
$\Delta V_{rms} = 0.707 \Delta V_{max} =$
$(0.707)(215 \text{ V}) = \boxed{152 \text{ V}}$

9. b

Given
$\Delta V_1 = 4850$ V
$N_1 = 2500$ turns
$N_2 = 5.0 \times 10^1$ turns

Solution

$$\Delta V_2 = \Delta V_1\frac{N_2}{N_1} =$$

$$(4850 \text{ V})\left(\frac{5.0 \times 10^1 \text{ turns}}{2500 \text{ turns}}\right) = \boxed{97 \text{ V}}$$

10. d
11. a
12. b
13. d
14. When an applied magnetic field approaches the coil of wire, the direction of the induced current produces an induced magnetic field that is in a direction opposite the approaching (or strengthening) magnetic field. As a result, these magnetic fields repel each other. When the applied magnetic field moves away from the coil of wire, the direction of the induced current once again produces a magnetic field that is in a direction opposite the receding (or weakening) magnetic field. This time, however, the induced magnetic field is in the same direction as the receding magnetic field. As a result, these magnetic fields attract each other.

15. Back emf is an induced emf in a motor's rotating coil. It decreases motor efficiency, since the back emf reduces the net supply current available in the motor's coil.

16. Electromagnetic waves exhibit both wave and particle behavior depending on the wave's frequency (or wavelength). When an electromagnetic wave behaves more like a stream of particles, the "particles" are called photons. A photon is a particle that carries energy but has zero rest mass. A high-energy photon behaves more like a particle, while a low-energy photon behaves more like a wave.

17. Answers may vary. Sample answer: Radio waves work well for transmitting information across long distances because the long wavelengths can easily travel around objects. Radio waves help scientists understand deep space objects because the long wavelengths of radio waves pass through Earth's atmosphere.

18. -3.2 A

Given

$N = 40.0$ turns
$A = 0.50$ m^2
$B_i = 0.00$ T
$B_f = 0.95$ T
$\Delta t = 2.0$ s
$\theta = 0.00°$
$R = 3.0\ \Omega$

Solution

Use Faraday's law of magnetic induction to calculate emf.

$$\text{emf} = -N\frac{\Delta\Phi_M}{\Delta t} = -N\frac{\Delta ABcos\theta}{\Delta t} =$$

$$-NAcos\theta\frac{\Delta B}{\Delta t} = -NAcos\theta\frac{(B_f - B_i)}{\Delta t}$$

$$= -(40.0 \text{ turns})(0.50 \text{ m}^2)$$

$$(cos\ 0.00°)\frac{(0.95 \text{ T} - 0.00 \text{ T})}{(2.0 \text{ s})} =$$

$$-9.5 \text{ V}$$

Substitute the induced emf into the definition of resistance to determine the induced current in the coil.

$$I = \frac{\text{emf}}{R} = \frac{(-9.5 \text{ V})}{(3.0\ \Omega)} = \boxed{-3.2 \text{ A}}$$

19. 9.8×10^2 V

Given

$\Delta V_{rms} = 6.9 \times 10^2$ V

Solution

Rearrange the equation, $\Delta V_{rms} = 0.707\Delta V_{max}$, to solve for ΔV_{max}.

$$\Delta V_{max} = \frac{\Delta V_{rms}}{0.707} = \frac{(6.9 \times 10^2 \text{ V})}{(0.707)} =$$

$$\boxed{9.8 \times 10^2 \text{ V}}$$

20. -2370 V

Given

$N_1 = 196$ turns
$N_2 = 9691$ turns
$A = 0.180$ m^2
$B_i = 0.000$ T
$B_f = 0.950$ T
$\Delta t = 0.700$ s
$\theta = 0.000°$

Solution

Use Faraday's law of magnetic induction to calculate emf.

$$\text{emf} = -N\frac{\Delta\Phi_M}{\Delta t} = -N\frac{\Delta ABcos\theta}{\Delta t} =$$

$$-NAcos\theta\frac{\Delta B}{\Delta t} = -NAcos\theta\frac{(B_f - B_i)}{\Delta t}$$

$$= -(196 \text{ turns})(0.180 \text{ m}^2)$$

$$(cos\ 0.000°)\frac{(0.950 \text{ T} - 0.00 \text{ T})}{(0.700 \text{ s})}$$

$$= -47.9 \text{ V}$$

Use the transformer equation to find the induced emf in the secondary coil.

$$\Delta V_{2rms} = \Delta V_{1rms}\frac{N_2}{N_1} =$$

$$(-47.9 \text{ V})\left(\frac{9691 \text{ turns}}{196 \text{ turns}}\right) = \boxed{-2370 \text{ V}}$$

Atomic Physics

CHAPTER TEST A (GENERAL)

1. a	**10.** a
2. d	**11.** c
3. b	**12.** a
4. a	**13.** b
5. c	**14.** a
6. d	**15.** b
7. c	**16.** d
8. a	**17.** b
9. d	**18.** d

19. An emission spectrum is a unique series of spectral lines emitted by an atomic gas when a potential difference is applied across the gas.
20. the wave model
21. the particle model
22. The energy of the incoming photons is equal to twice the maximum kinetic energy of the emitted photoelectrons.
23. 1.2 eV

Given
$f = 3.0 \times 10^{14}$ Hz
$h = 6.63 \times 10^{-34}$ J•s

Solution
$E = hf$
$E = (6.63 \times 10^{-34} \text{ J•s})(3.0 \times 10^{14} \text{ Hz})$
$\left(\dfrac{1 \text{ eV}}{1.60 \times 10^{-19} \text{ J}} \right)$
$E = \boxed{1.2 \text{ eV}}$

24. 1.13 eV

Given
$E_6 = -0.378$ eV
$E_3 = -1.51$ eV

Solution
$E = E_{initial} - E_{final} = E_6 - E_3$
$E = -0.378 \text{ eV} - (-1.51 \text{ eV}) = \boxed{1.13 \text{ eV}}$

25. 3.1×10^{-10} m, or 0.31 nm

Given
$m = 1.67 \times 10^{-27}$ kg
$v = 1.3 \times 10^3$ m/s
$h = 6.63 \times 10^{-34}$ J•s

Solution
$\lambda = \dfrac{h}{p} = \dfrac{h}{mv}$

$\lambda = \dfrac{(6.63 \times 10^{-34} \text{ J•s})}{(1.67 \times 10^{-27} \text{ kg})(1.3 \times 10^3 \text{ m/s})}$

$= 3.1 \times 10^{-10} \text{ m} = \boxed{0.31 \text{ nm}}$

Atomic Physics

CHAPTER TEST B (ADVANCED)

1. c

Solution
$E = hf$
$f = \dfrac{E}{h} = \dfrac{1.99 \times 10^{-19} \text{ J}}{6.63 \times 10^{-34} \text{ J•s}}$
$E = \boxed{3.00 \times 10^{14} \text{ Hz}}$

2. b
3. c

Solution
$KE_{max} = hf - hf_t$
$KE_{max} = 3.0 \text{ eV} - 1.6 \text{ eV} = \boxed{1.4 \text{ eV}}$

4. b
5. c
6. c
7. b
8. c
9. Planck proposed that resonators could only absorb and then reemit discrete amounts of light energy called quanta.
10. The constantly accelerated electrons in Rutherford's model of the atom would continuously radiate electromagnetic waves, and therefore would be unstable. Also, his model did not explain spectral lines.
11. The resulting spectrum is an absorption spectrum, which appears as a nearly continuous spectrum with dark lines where light of given wavelengths is absorbed by the gases in the cloud.
12. The transitions from any of the excited energy levels to the ground state will produce photons with the greatest energy, and therefore the shortest wavelengths.
13. Earth's magnetic field draws charged particles from the sun toward the poles, where the particles collide with atoms in Earth's atmosphere. These atoms give up the energy acquired in the collisions as spontaneous emission of photons, producing an aurora. Because there are more collisions near the poles, more light is emitted, producing a brighter aurora more often.
14. Light of short wavelengths is better. Momentum transfer is most easily observed in particle collisions, and photons that have shorter wavelengths behave more like particles than do photons with longer wavelengths.

15. The standing wave is stable because there is no interference between the incident and reflected wave. This occurs when there are an integral number of wavelengths for the wave along the string. This is comparable to electrons in a Bohr atom, which only have stable orbits when there are an integral number of electron wavelengths along the path of the orbit.

16. The interaction of light and microscopic particles is so slight that the position and the momentum of a large object are imperceptibly affected, whereas the effect of interactions between light or microscopic particles and other microscopic particles can be relatively large.

17. 2.1 eV

Given

$f = 5.0 \times 10^{14}$ Hz

$h = 6.63 \times 10^{-34}$ J•s

Solution

$E = hf$

$E = (6.63 \times 10^{-34} \text{ J•s})(5.0 \times 10^{14} \text{ Hz})$

$\left(\dfrac{1 \text{ eV}}{1.60 \times 10^{-19} \text{ J}} \right)$

$E = \boxed{2.1 \text{ eV}}$

18. 3.98 eV

Given

$\lambda = 312$ nm

$h = 6.63 \times 10^{-34}$ J•s

$c = 3.00 \times 10^{8}$ m/s

Solution

$E = hf$

$c = f\lambda$

$E = \dfrac{hc}{\lambda} =$

$\dfrac{(6.63 \times 10^{-34} \text{ J•s})(3.00 \times 10^{8} \text{ m/s})}{312 \text{ nm}}$

$\left(\dfrac{10^{9} \text{ nm}}{1 \text{ m}} \right)\left(\dfrac{1 \text{ eV}}{1.60 \times 10^{-19} \text{ J}} \right)$

$E = \boxed{3.98 \text{ eV}}$

19. 2.73×10^{14} Hz

Given

$E_6 = -0.378$ eV

$E_3 = -1.51$ eV

$h = 6.63 \times 10^{-34}$ J•s

Solution

$E = hf$

$E = E_{initial} - E_{final} = E_6 - E_3$

$f = \dfrac{E_6 - E_3}{h} =$

$\dfrac{(-0.378 \text{ eV} - (-1.51 \text{ eV}))}{6.63 \times 10^{-34} \text{ J•s}}$

$\left(\dfrac{1.60 \times 10^{-19} \text{ J}}{1 \text{ eV}} \right) =$

$\left(\dfrac{1.13 \text{ eV}}{6.63 \times 10^{-34} \text{ J•s}} \right)\left(\dfrac{1.60 \times 10^{-19} \text{ J}}{1 \text{ eV}} \right)$

$f = \boxed{2.73 \times 10^{14} \text{ Hz}}$

20. 1.5×10^{-12} m, or 1.5 pm

Given

$m = 1.67 \times 10^{-27}$ kg

$v = 2.7 \times 10^{5}$ m/s

$h = 6.63 \times 10^{-34}$ J•s

Solution

$\lambda = \dfrac{h}{p} = \dfrac{h}{mv}$

$\lambda = \dfrac{(6.63 \times 10^{-34} \text{ J•s})}{(1.67 \times 10^{-27} \text{ kg})(2.7 \times 10^{5} \text{ m/s})}$

$= 1.5 \times 10^{-12} \text{ m} = \boxed{1.5 \text{ pm}}$

Subatomic Physics

CHAPTER TEST A (GENERAL)

1. d	**6.** b
2. a	**7.** b
3. c	**8.** c
4. d	**9.** b
5. a	
10. a	

Solution

Mass number of X = 240 − 236 = 4

Atomic number of X = 94 − 92 = 2

The emitted particle is a helium-4 nucleus, or an alpha particle.

11. b

Solution

Mass number of X = 226 − 222 = 4
Atomic number of X = 88 − 86 = 2
From the periodic table, the nucleus
with an atomic number of 2 is He, so
helium-4 nuclei (alpha particles) form
the unknown reaction products.

12. c

13. c

14. a

15. a

16. b

17. d

18. a

19. d

20. b

21. Half-life is the time required for half
the original nuclei of a radioactive
material to undergo radioactive decay.

22. Nuclear fission is a process during
which a heavy nucleus splits into two
or more lighter nuclei.

23. The four fundamental interactions of
physics operated in a unified manner.
The high temperatures and energy
caused all particles and energy to be
indistinguishable.

24. gravitational, weak, electromagnetic,
strong

25. 492.26 MeV

Given

Z of $^{56}_{26}$Fe = 26

N of $^{56}_{26}$Fe = 56 − 26 = 30

atomic mass of $^1_1$H = 1.007 825 u

m_n = 1.008 665 u

atomic mass of $^{56}_{26}$Fe = 55.934 940 u

c^2 = 931.49 MeV/u

Solution

$\Delta m = Z$(atomic mass of $^1_1$H) +
$\quad Nm_n$ − atomic mass
$\quad = (26)(1.007\ 825\ u) +$
$\quad (30)(1.008\ 665\ u) - 55.934\ 940\ u$
$\quad = 0.528\ 460\ u$

$E_{bind} = (0.528\ 460\ u)(931.49\ MeV/u) =$
$\boxed{492.26\ MeV}$

Subatomic Physics

CHAPTER TEST B (ADVANCED)

1. b

Solution

number of protons in Pb = number of
protons in Pb-210 = 210 − 128 = 82
number of neutrons in Pb-206 =
$\quad$ 206 − 82 = 124

2. c

3. c

4. d

Solution

Mass number of X = 200 − 216 = 4
Atomic number of X = 86 − 84 = 2
From the periodic table, the nucleus
with an atomic number of 2 is He.

5. b

6. c

Solution

12.5 percent = 0.125 = 1\8 = 1\23
The substance undergoes 3 half-lives.

7. b

8. a

9. c

10. gamma particles, beta particles, alpha
particles

11. $^4_2$He

Solution

Mass number of unknown =
$\quad$ 230 − 226 = 4
Atomic number of unknown =
$\quad$ 90 − 88 = 2
From the periodic table, the nucleus
with an atomic number of 2 is He.

12. neutron

13. Reactor fuels must be processed or
enriched to increase the proportion of
uranium-235, which undergoes fission
and releases energy, to a level that will
sustain the reaction.

14. The atomic number indicates the total
number of protons in the nucleus,
while the mass number indicates the
total number of nucleons (protons and
neutrons) in the nucleus. The neutron
number is the difference between
these numbers.

15. Leptons appear to be fundamental, meaning that they do not break down to smaller particles. Hadrons are particles that are made from combinations of smaller particles called quarks.

16. 333.72 MeV

Given

Z of $^{39}_{19}$K = 19

N of $^{39}_{19}$K = 39 − 19 = 20

atomic mass of $^{1}_{1}$H = 1.007 825 u

m_n = 1.008 665 u

atomic mass of $^{39}_{19}$K = 38.963 708 u

c^2 = 931.49 MeV/u

Solution

Δm = Z(atomic mass of $^{1}_{1}$H) + Nm_n − atomic mass

= (19)(1.007 825 u) + (20)(1.008 665 u) − 38.963 708 u

= 0.358 267 u

E_{bind} = (0.358 267 u)(931.49 MeV/u) = $\boxed{333.72 \text{ MeV}}$

17. 7.9157 MeV/nucleon

Given

Z of $^{197}_{79}$Au = 79

N of $^{197}_{79}$Au = 197 − 79 = 118

atomic mass of $^{1}_{1}$H = 1.007 825 u

m_n = 1.008 665 u

atomic mass of $^{197}_{79}$Au = 196.966 543 u

c^2 = 931.49 MeV/u

Solution

Δm = Z(atomic mass of $^{1}_{1}$H) + Nm_n − atomic mass

= (79)(1.007 825 u) + (118)(1.008 665 u) − 196.966 543 u

= 1.674 102 u

E_{bind} = (1.674 102 u)(931.49 MeV/u) = 1559.4 MeV

E_{bind}/n = $\dfrac{1559.4 \text{ MeV}}{197 \text{ nucleons}}$ =

$\boxed{7.9157 \text{ MeV/nucleon}}$

18. 2.0 h

Given

initial activity = 800.0 counts/s

final activity = 200.0 counts/s

t = 4.0 h

Solution

t = 4.0 h

amount of sample remaining =

$\dfrac{200 \text{ counts/s}}{800.0 \text{ counts/s}}$ = 0.2500

$0.2500 = (0.5)^2$, so 2 half-lives have passed.

$T_{1/2} = \dfrac{1}{2}t = \dfrac{1}{2}(4.0 \text{ h}) = \boxed{2.0 \text{ h}}$

19. 36.9 years

Given

$T_{1/2}$ = 12.3 years

percentage of hydrogen-3 remaining after decay = 12.5

Solution

$0.125 = \dfrac{1}{8.00} = \left(\dfrac{1}{2}\right)^3$

It takes 12.3 years for $\frac{1}{2}$ the sample to decay. Therefore, the sample decays to

$\dfrac{1}{8} = \left(\dfrac{1}{2}\right)^3$ of its original strength in

3(12.3 years) = $\boxed{36.9 \text{ years}}$.

20. 5.5×10^{14} Bq

Given

fraction of iodine-131 decayed = $\dfrac{3}{4}$

t = 16.14 days

N = number of iodine-131 nuclei = 5.5×10^{20}

Solution

If $\dfrac{3}{4}$ of the sample decays in a given time, then $\dfrac{1}{4}$ of the sample remains.

$\dfrac{1}{4} = \left(\dfrac{1}{2}\right)^2$

The sample decays to $\dfrac{1}{4} = \left(\dfrac{1}{2}\right)^2$ of its original strength in 16.14 days = $2T_{1/2}$.

$T_{1/2} = \dfrac{16.14 \text{ days}}{2}$ = 8.070 days

$T_{1/2} = \dfrac{0.693}{\lambda}$

$\lambda = \dfrac{0.693}{T_{1/2}}$ =

$\dfrac{0.693}{(8.070 \text{ days})(24 \text{ h/day})(3600 \text{ s/h})}$

= $9.94 \times 10^{-7} \text{ s}^{-1}$

activity = λN =

$(9.94 \times 10^{-7} \text{ s}-1)(5.5 \times 10^{20} \text{ decays})$

= 5.5×10^{14} decays/s =

$\boxed{5.5 \times 10^{14} \text{ Bq}}$

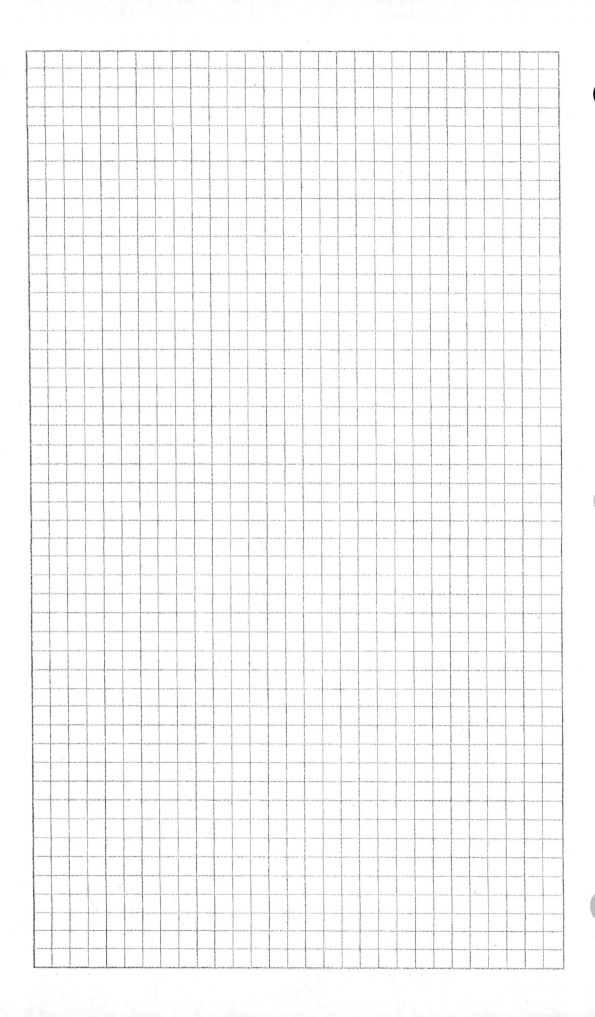

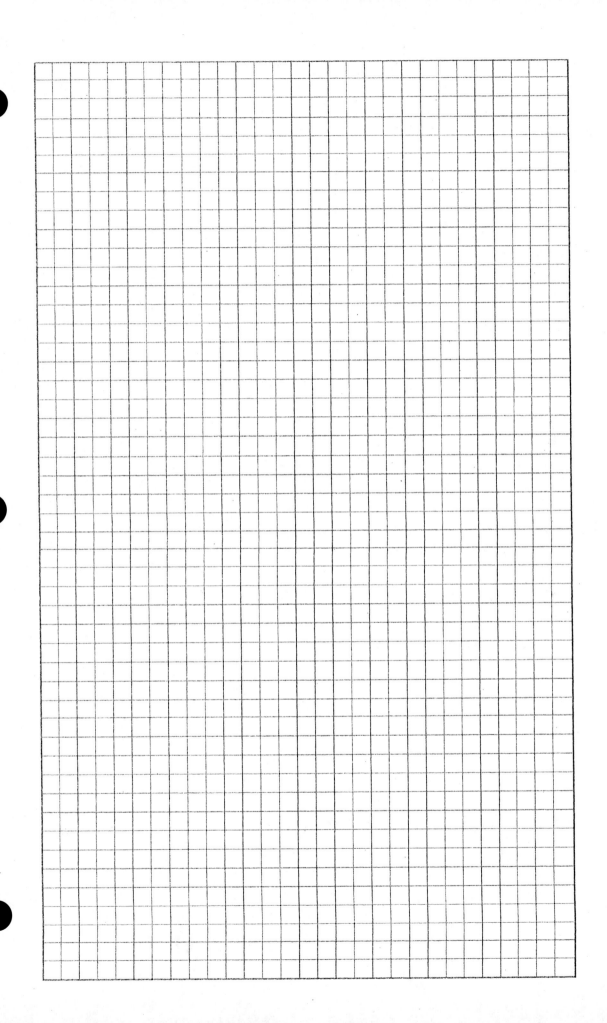

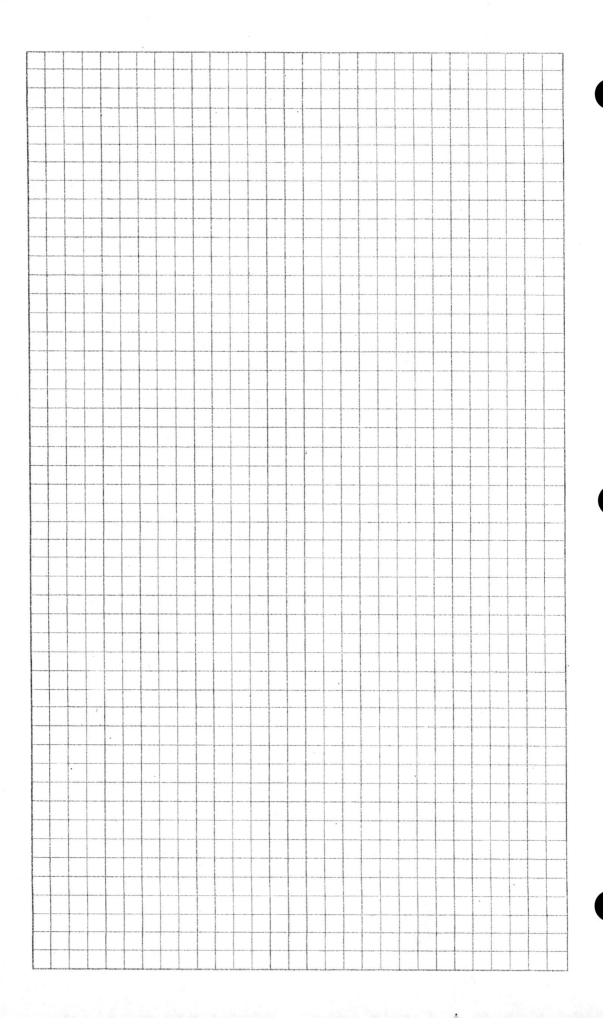

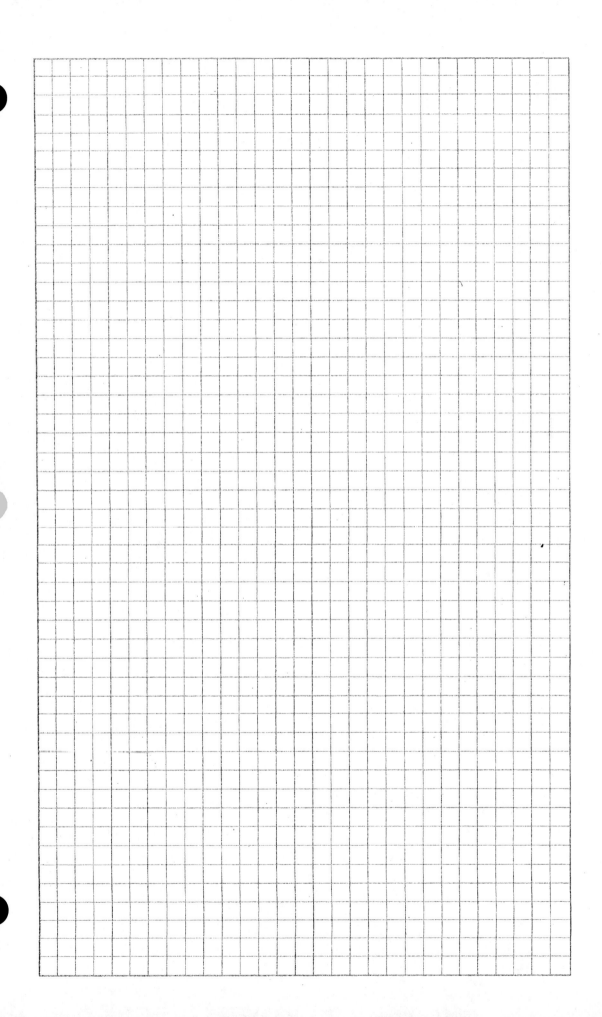

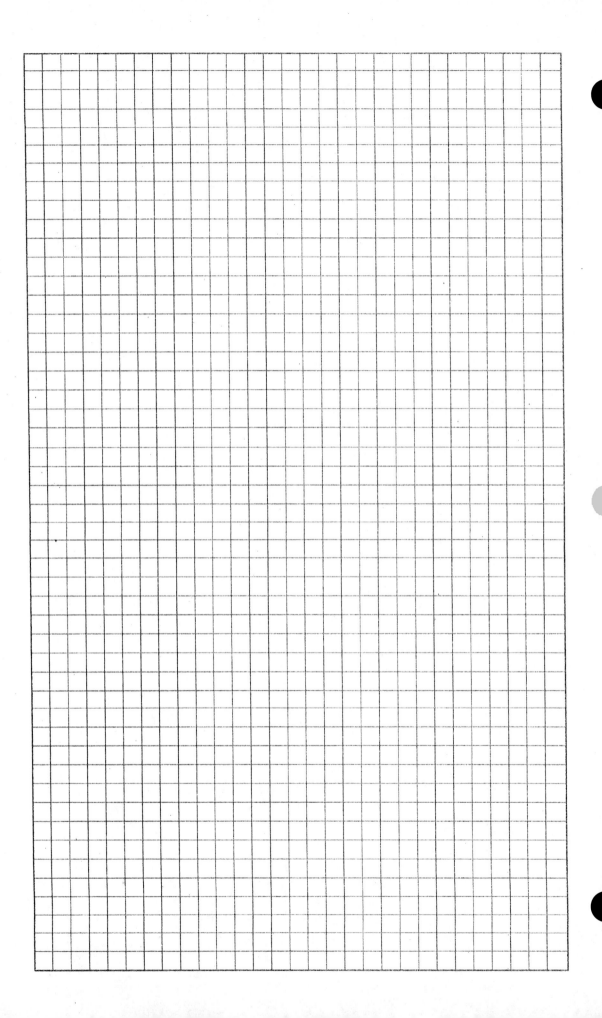

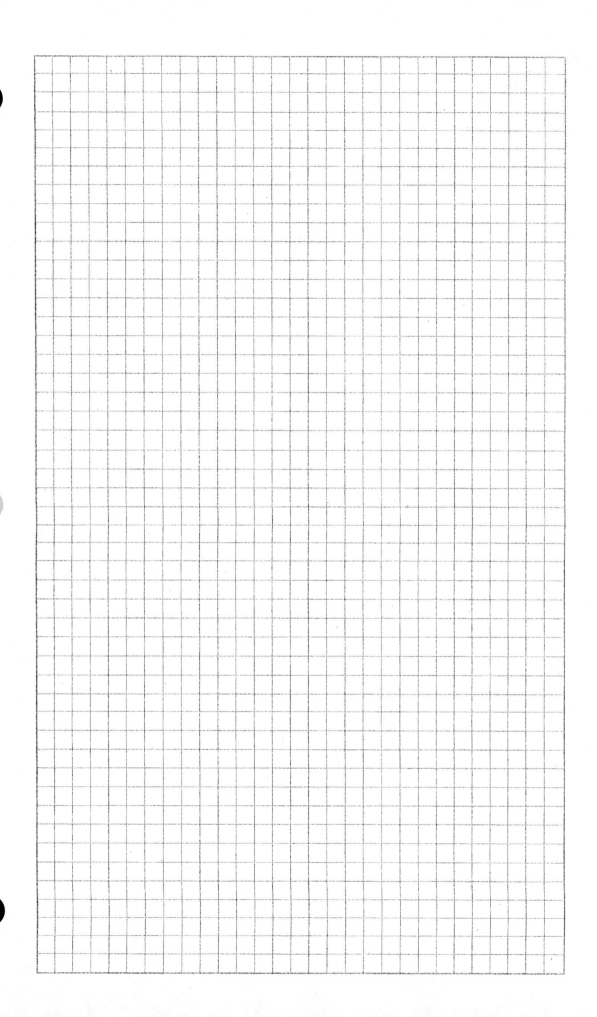

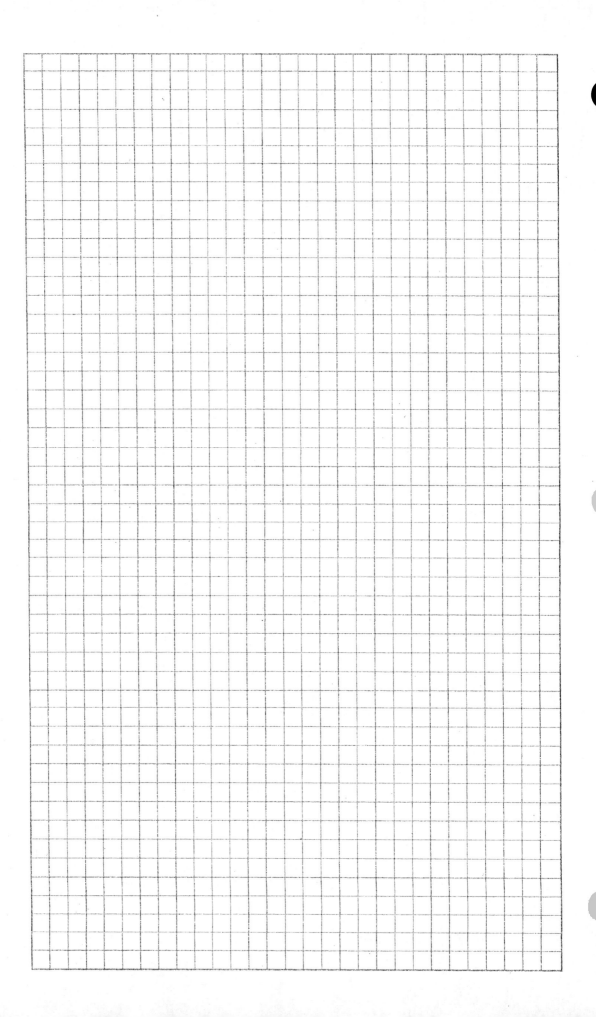